Introduction
to
Statistical Methods
Volume II

Introduction
to
Statistical Methods
Volume II

Jagdish S. Rustagi

Rowman & Allanheld

PUBLISHERS

ROWMAN & ALLANHELD

Published in the United States of America in 1985
by Rowman & Allanheld, Publishers
(A division of Littlefield, Adams & Company)
81 Adams Drive, Totowa, New Jersey 07512

Library of Congress Cataloging in Publication Data
(Revised for v. 2)

Rustagi, Jagdish S.
  Introduction to statistical methods.

  Includes indexes.
  1. Statistics. I. Title.
  QA276.12.R87   1985   519.5   84—6911
  ISBN 0—86598—127—2 (v. 1)
  ISBN 0—86598—128—0 (v. 2))

84 85 86 /10 9 8 7 6 5 4 3 2 1

Printed in the United States of America

Photoset at Thomson Press (India) Limited, New Delhi

To Arjun

Contents of Volume II

Contents of Volume I

List of Figures

List of Tables

Appendix Tables

Preface

Statistical methods are being used by more and more scientists, medical clinicians, government administrators, business managers, industrial engineers, and market researchers. Statistics now plays a major role in policy decisions, quality assurance, public health, and clinical trials. Students in the behavioral sciences, biomedical sciences, agricultural sciences, and business administration are required to take statistics courses. With easy access to pocket calculators and microcomputers, statistical computations have become painless.

This book provides an introduction to statistical methods. Volume I covers the basic topics in probability, statistical inference, and regression analysis. It also includes a brief introduction to the design and analysis of experiments. Volume II covers survey sampling, quality assurance, survival analysis, statistical bioassay, and sequential experimentation. Topics on clinical trials, life tables, and compartmental analysis are also included.

The mathematical background required for this book is high-school algebra. Derivations of results are not given. Statistical concepts are explained with as little mathematics as possible. Examples from real-life situations are given wherever possible. Simple examples with artificial data are used to illustrate many statistical concepts. Numerical computations are not stressed. Statistical packages such as SAS, SPSS, and BMDP contain almost all the standard statistical techniques and are now used in classrooms. Instructors are encouraged to use such packages to further illustrate a statistical notion.

The purpose of the book is to introduce the reader to statistical philosophy and standard methods in statistics. For the use of statistical techniques in complicated situations, a statistician should be consulted. Volume I can be covered in a two-quarter course meeting three times a week. Some topics need to be omitted if the course is to be covered in a semester. Topics from Volume II can be covered in one semester.

Acknowledgments

I am grateful to several generations of students at Carnegie Institute of Technology, Michigan State University, Aligarh Muslim University, University of Cincinnati College of Medicine, and The Ohio State University, where I have taught. I am highly obliged to the authors of the textbooks and journal articles who are quoted throughout the book. Professor Paul Switzer was a source of constant encouragement during the preparation of the manuscript, and I am very grateful to him. Professor Abhaya Indrayan provided extensive suggestions and corrections on the first draft. His help, in fact, amounts almost to coauthorship of the book. This volume was improved considerably as a result of a critical review of Professor Michael A. Schork, who made many suggestions and comments.

Several important suggestions and comments were made on various chapters by Professors Prem Puri, H. N. Nagaraja, Prem Goel and Om P. Aggarwal. I am deeply grateful to them. The manuscript was completed during a sabbatical from The Ohio State University spent at Stanford University and the Indian Statistical Institute, Delhi Center, and I am obliged to both.

For the excellent manuscript typing, I thank Bach Hong, Dolores Wills, and Mona Swanger. Professor Herman Chernoff has always been a source of inspiration and encouragement and I am grateful to him. I thank my wife, Kamla, and my family for providing a peaceful environment during the writing of the book. I am grateful to senior editor Paul Lee, of Littlefield, Adams & Company, who has been very helpful in various aspects of publication matters.

I shall be greatly obliged to readers for their suggestions and comments.

Introduction
to
Statistical Methods
Volume II

chapter one

Sample Surveys

In health-related areas, data are routinely collected on large number of items for policy and decision making. Many problems arising in the organization of community health programs, as well as in the administration of clinical trials, belong to the areas of sample surveys. Surveys are used routinely in the agricultural, social, behavioral, and administrative sciences. Many decisions in large organizations are based on data collected by sample survey methods. National governments and the United Nations also conduct periodic surveys. National surveys of cattle on feed to predict milk and meat production, labor statistics to predict the rate of unemployment, and many other surveys are conducted periodically by the federal government.

Surveys of health facilities, health manpower resources, infectious diseases, and many other surveys related to public health are the concern of the National Center of Health Statistics in the United States. In the National Health Survey Act of 1956, Congress authorized a continuing program of health surveys to be carried out by the Public Health Service to provide reliable and comprehensive information for administrators, research, and the public. Many surveys such as the national fertility survey, the health examination survey, and perodontal disease surveys have been completed recently.

The U.S. Bureau of Census began a survey in October 1983 to gather data on all sources of income, household net worth, taxes, labor force participation, and other social and economic characteristics from 18,000 households. The purpose of the survey, among other things, is to find the number of people and families who are eligible for benefits but not getting them, whether federal help is distributed equitably under present programs and rules, and the effect of changes in tax policy on program participation and income.

Survey methods are used by auditors to check the accuracy of accounts, by marketing researchers to find potential markets for certain products as well as consumer interest in a product, and by political scientists for gauging the opinions of the electorate.

This chapter discusses elementary concepts of sample surveys. The population surveyed is considered finite. One object of any survey is to estimate the *total* or *mean* of a character of interest in the population. Complete enumeration of the whole population is not only costly and time-consuming, but the results are not always accurate. Random sampling from the population provides the estimates for the population parameters. Various methods of sampling and estimation procedures are described. The statistical techniques needed to answer various questions in a survey have been discussed in Volume 1; the design aspects are emphasized here.

Another area of sampling is concerned with moving populations, such as estimating the total amount of wildlife in a forest or the total number of fish in a lake. The survey designs of moving populations are different, however, and are not discussed here.

Simple Random Sampling

We assume that the population to be surveyed has N units, such as households, and $y_1, y_2, \ldots y_N$ are the values of certain characteristics of the units, such as income. The *population total* is denoted by:

$$Y = y_1 + y_2 + \cdots + y_N \tag{1.1}$$

The *population mean* is denoted by:

$$\bar{Y} = \frac{1}{N} Y \tag{1.2}$$

The *population variance* is:

$$S^2 = \sum (y_i - \bar{Y})^2 / (N - 1) \tag{1.3}$$

Y and S^2 are parameters of the finite population. Ordinarily, the parameters of a population are denoted by Greek letters. In finite population sampling, however, they are denoted by English capital letters. The estimates of the parameters will be denoted with a carat ($\hat{\ }$).

A random sample of n units can be chosen from the population in one of two ways:

   (i) Sampling without replacement
  (ii) Sampling with replacement

Under (i), *simple random sampling* requires the selection of n units with probability $1/\binom{N}{n}$, since the total number of possible samples is $\binom{N}{n}$.

As discussed in Chapter 5, Volume I, a simple random sample can be obtained by selecting one unit at a time without replacement, so as to give equal probabilities of selection to each unit at a new selection. The use of random-numbers tables for obtaining a random sample was also discussed. Under (ii), simple random sampling requires the selection of each unit in the sample with probability $1/N$ at every selection.

Notice that the randomness in the finite population is introduced by the process of obtaining a sample. We denote the sample mean by $\bar{y}$ and the sample variance by $s^2 = \sum\limits_{i=1}^{n} (y_i - \bar{y})^2/(n-1)$

Whether sampling is with replacement or without replacement, the expected value of $\bar{y}$ is $\bar{Y}$. That is,

$$E(\bar{y}) = \bar{Y} \tag{1.4}$$

For sampling without replacement,

$$\text{Var}(\bar{y}) = \left(1 - \frac{n}{N}\right)\frac{S^2}{n} \tag{1.5}$$

The quantity $f = \dfrac{n}{N}$ is called the *sampling fraction*, and $1-f$ is called the *population correction.*

In the infinite population sampling, the variance of the sample mean is the variance of the population divided by n. However, in the finite population sampling, the variance of the sample mean is adjusted by the finite population correction $1-f$. When N is large with respect to n, f can be neglected and can be equated as zero. In this section, we will restrict ourselves to sampling without replacement.

The unbiased estimates of the average and variance are given by

$$\hat{\bar{Y}} = \bar{y} \tag{1.6}$$

and

$$\hat{S}^2 = s^2 \tag{1.7}$$

Using the above, the variance of $\bar{y}$ is estimated by:

$$\hat{V}(\bar{y}) = (1 - f)s^2/n \tag{1.8}$$

When n is large, using the Central Limit Theorem, the probability distribution of $\bar{y}$ is normal with mean $\bar{Y}$ and variance $(1 - f)S^2/n$. This fact is used to obtain *confidence interval* estimates for $\bar{Y}$.

The $(1 - \alpha)$-level confidence interval for $\bar{Y}$ is

$$\left(\bar{y} - z_{1-\alpha/2}S\sqrt{\frac{1-f}{n}}, \quad \bar{y} + z_{1-\alpha/2}S\sqrt{\frac{1-f}{n}}\right) \tag{1.9}$$

where $z_{1-\alpha/2}$ is the $(1-\alpha/2)$-th percentile of the normal distribution. If S is to be estimated by s when n is small, we can use the t-distribution with $n-1$ degrees of freedom in place of a normal distribution.

The $(1-\alpha)$-level confidence interval for $\bar{Y}$ when S is not known is:

$$\left(\bar{y} - t_{n-1,1-\alpha/2}\, s\, \sqrt{\frac{1-f}{n}}, \quad \bar{y} + t_{n-1,1-\alpha/2}\, s\, \sqrt{\frac{1-f}{n}} \right) \tag{1.10}$$

Selection of a Simple Random Sample

As described in Chapter 6, Volume I, the table of random numbers is used to select a random sample. Recall that the population units are numbered from 1 to N. From Table VII of random numbers, we choose n distinct numbers; then we select units with those numbers.

Example 1.1: In a junior school with 300 students, a survey is made to find the average number of cavities in teeth per student. A random sample of 10 students is obtained and the number of cavities are:

$$2, \quad 1, \quad 3, \quad 0, \quad 5, \quad 2, \quad 1, \quad 3, \quad 2, \quad 1$$

Here $N = 300$, $n = 10$, $\bar{y} = 20/10 = 2$

$$s^2 = 2$$

$$f = \frac{10}{300} = .0333$$

$$\hat{V}(\bar{y}) = (1 - .0333) \times \frac{2}{10} = .1933$$

The .95 confidence interval for the population mean $\bar{Y}$ is given by:

$$(2 - t_{9,.975}\sqrt{.1933}, \quad 2 + t_{9,.975}\sqrt{.1933})$$
$$= (2 - 2.262\sqrt{.1933}, \quad 2 + 2.262\sqrt{.1933})$$
$$= (1, 3)$$

Sampling to Estimate a Proportion

Suppose we want to estimate a proportion such as the proportion of Medicare patients admitted to a hospital, or the proportion of minority workers in a large corporation, or the proportion of smokers among university students. Using a random sample, we can estimate the proportion in the same way that we estimate the average. The observations $y_1, y_2, \ldots y_n$ are now replaced by ones and zeros—one representing the presence of the characteristic and zero representing its absence. The population proportion P, then, is estimated by the sample proportion p. That is,

$$\hat{P} = p \tag{1.11}$$

and p is an unbiased estimate of P.

The sample proportion p has the variance

$$V(p) = \frac{PQ}{n}\left(\frac{N-n}{N-1}\right) \tag{1.12}$$

and an unbiased estimate for $V(p)$ is

$$s_p^2 = \hat{V}(p) = \frac{pq}{n-1}\left(\frac{N-n}{N}\right) = \frac{pq}{n-1}(1-f) \tag{1.13}$$

The confidence intervals for P can be obtained as in (1.9) when n is large. We have $(1-\alpha)$-level confidence interval for P given by the limits:

$$p \pm \left[z_{1-\alpha/2}\sqrt{\frac{(1-f)pq}{n-1}} + \frac{1}{2n} \right] \tag{1.14}$$

Here $\dfrac{1}{2n}$ is a correction for continuity.

Example 1.2: A random sample of 25 students is taken from students in a high school with 2,500 students. It is found that 10 students smoke cigarettes. To find the estimate of proportion of smokers in the high school, we have:

$$p = \frac{10}{25} = .4$$

Since $n = 25$ and $N = 2,500$, we have the estimate of its variance:

$$s_p^2 = \frac{.4 \times .6}{24}\left(1 - \frac{25}{2500}\right) = \frac{.4 \times .6 \times .99}{24}$$
$$= .0099$$

The 95% confidence-interval estimate of P is:

$$\left[.4 - \left(1.96\sqrt{.0099} + \frac{1}{50}\right) \quad .4 + \left(1.96\sqrt{.0099} + \frac{1}{50}\right) \right]$$
$$= [.185, \quad .615]$$

Exercises

1. A simple random sample of 12 households is taken from a population consisting of 300 households, and their monthly income is given below (in thousands of dollars). Estimate the average income and its variance per

household. What is the estimate of the total income for the population?

$$3, \quad 2.5, \quad 7.5, \quad 4.3, \quad 2.3, \quad 5, \quad 7, \quad 9, \quad 10, \quad 8, \quad 4, \quad 5.5$$

2. The population of patients in a hospital is studied for their length of stay. There are a total of 250 patients, and a random sample of 9 patients gives the data:

$$3, \quad 7, \quad 5, \quad 1, \quad 2, \quad 8, \quad 15, \quad 17, \quad 6$$

Obtain the mean length of stay and its variance. Find a 95% confidence interval for the mean stay.

3. The population in a village of 5,000 persons was screened for the presence of AIDS. A random sample of 35 persons was taken and 5 cases of AIDS were found. Find the estimate of the proportion of the disease in the population and give its 90% confidence-interval estimate.

4. To estimate the total retail sales in a small city, a random sample of 10 stores was made. Find the estimated total retail sales (in millions) and the 95% confidence interval for the population total. The sample provided the data:

$$1.6, \quad 2.5, \quad 1.3, \quad 0.2, \quad 1.6, \quad 2.1, \quad 3.1, \quad .5, \quad 1.8, \quad .04$$

Stratified Random Sampling

When the population consists of units that are very different from each other, we may be able to divide the population into several homogeneous groups before sampling called *strata*. The process of stratifying the population allows a fair representation of various kinds of items in the population, since the random sampling will be done in each stratum, rather than over the whole population. For example, if we want to estimate the average length of stay of patients in a hospital, we can group the patients in various categories (strata) according to the kinds of diseases. We know that the duration of stay differs a great deal from disease to disease. The patients will be divided among various strata and a simple random sample in each stratum will be taken. This sampling procedure is known as *stratified random sampling*.

The strata are chosen so that there is very little variability among the units within the stratum. The choice of strata in sample survey design is similar to the choice of blocks in an experiment-design context.

Suppose the population is divided into L strata. Let the number of units in hth stratum be N_h and let $N_1 + N_2 + \cdots + N_L = N$. We choose a random sample of size n_h from the hth stratum with $n = n_1 + n_2 + \cdots + n_L$.

Let $\bar{y}_h$ be the sample mean of the hth stratum. Then the estimate of the population mean $\bar{Y}$ is given by:

$$\bar{y}_{st} = \sum_{h=1}^{L} \frac{N_h \bar{y}_h}{N} \tag{1.15}$$

Let:

$$W_h = \frac{N_h}{N}$$

Then

$$\bar{y}_{st} = \sum_{h=1}^{L} W_h \bar{y}_h$$

can be expressed as a weighted average of the strata averages, and the weights are proportions of the number of units in the strata. Note that the weights add to one.

This estimate is an unbiased estimate of the population mean, since:

$$E(\bar{y}_{st}) = \sum_{h=1}^{L} W_h E(\bar{y}_h) = \sum_{h=1}^{N} W_h \bar{Y}_h = \bar{Y}.$$

We have used the fact that the expected value of a linear combination of random variables is the linear combination of the expected values.

In the case of $\frac{N_h}{N} = \frac{n_h}{n}$, the process of simple random sampling and stratified random sampling provide the same estimates. This allocation procedure is known as *proportional allocation*. The assignment of an observation to the sample under proportional allocation is such that the proportion of observations from a stratum is the same as the proportion of the total number of observations in the stratum with the population.

Now the variance of the average is given by

$$V(\bar{y}_{st}) = \frac{1}{N^2} \sum_{h=1}^{L} N_h(N_h - n_h)\frac{S_h^2}{n_h} \qquad (1.16)$$

where S_h^2 is given by

$$S_h^2 = \sum_{i=1}^{N_h} (y_{ih} - \bar{Y}_h)^2/(N_h - 1)$$

the variance of the hth stratum. The variance can be estimated by:

$$s_h^2 = \sum_{i=1}^{n_h} (y_{ih} - \bar{y}_h)^2/(n_h - 1) \qquad (1.17)$$

As a general rule, stratified random sampling gives a smaller variance of the estimates than simple random sampling. However, there are situations where this may not hold.

Example 1.3: Suppose in a county the farms are divided into strata according to their area. For reporting the yield of corn, a stratified random sample of 20 farms was taken based on proportional allocation. The mean yield per acre, the sample variances, and sample sizes are given below. $N_h =$ the number of farms in the hth stratum and $n_h =$ the number of farms in the sample selected with proportional allocation.

size	N_h	n_h	$\bar{y}_h$	s_h^2
very large	20	2	70	25
large	50	5	75	36
medium	80	8	60	25
small	50	5	65	25
Total	200	20		

The estimate of the mean is given by:

$$\bar{y}_{st} = \frac{1}{200}[(20 \times 70) + (50 \times 75) + (80 \times 60) + (50 \times 65)]$$

$$= 66$$

Here the weights are:

$$W_1 = \frac{20}{200} = .1$$

$$W_2 = \frac{50}{200} = .25$$

$$W_3 = \frac{80}{200} = .4$$

$$W_4 = \frac{50}{200} = .25$$

Notice that $W_1 + W_2 + W_3 + W_4 = 1$. The estimate of the variance is:

$$\hat{V}(\bar{y}_{st}) = \frac{1}{200^2}\left[\left(\frac{20 \times 18 \times 25}{2}\right) + \left(\frac{50 \times 45 \times 36}{5}\right)\right.$$

$$\left. + \left(\frac{80 \times 72 \times 25}{8}\right) + \left(\frac{50 \times 45 \times 25}{5}\right)\right]$$

$$= \frac{1}{40,000}[4500 + 16,200 + 18,000 + 11,250]$$

$$= 1.25$$

Estimating Proportions

The estimate for the population proportion using stratified random sampling is given by:

$$p_{st} = \frac{\sum_{h=1}^{L} N_h p_h}{N} \tag{1.18}$$

where p_h is the estimate of the proportion in the hth stratum. Its variance is given by

$$V(p_{st}) = \frac{1}{N^2} \sum_{h=1}^{L} \frac{N_h^2 (N_h - n_h)}{N_h - 1} \frac{P_h Q_h}{n_h} \tag{1.19}$$

where P_h is the stratum proportion for the hth stratum and $Q_h = 1 - P_h$. Its estimate is given by:

$$\hat{V}(p_{st}) = \frac{1}{N^2} \sum_{h=1}^{L} N_h (N_h - n_h) \frac{p_h q_h}{n_h - 1} \tag{1.20}$$

Sample Allocation

One of the important problems in stratified random sampling is how to allocate samples to various strata. A common method is to allocate n_h in such a way that the variance of the estimate is minimized for a fixed cost, or the cost of sampling is minimized for a fixed variance. The cost of sampling is generally taken to be linear. That is, there is some fixed overhead cost c_0 for the whole survey, and c_h is the cost per sampling unit in stratum h. The total cost is:

$$C = c_0 + \sum_{h=1}^{L} c_h n_h$$

It turns out that the optimum allocation is given by:

$$\frac{n_h}{n} = L \frac{N_h S_h / \sqrt{c_h}}{\sum_{h=1}^{L} (N_h S_h / \sqrt{c_h})} \tag{1.21}$$

That is, a larger sample is taken in a stratum if (i) it is cheaper to sample in the stratum, (ii) the variance within the stratum is large, or (iii) the size of the stratum is large.

When $c_h = c$—that is, in a stratum sample where every observation has the same cost:

$$n_h = n \frac{N_h S_h}{\sum N_h S_h} \tag{1.22}$$

The allocation in Equation (1.22) is called *Neyman allocation*. For details, see Cochran (1977).

Example 1.4: In an opinion survey, voters were divided into upper-, middle-, and low-income groups. The numbers in the three strata with the samples taken are given according to proportional allocation:

Income group	Size of stratum	Sample size (percent)
upper	150	5
middle	1,500	50
low	1,350	45

Suppose the proportion of approval of President Reagan's policies is given by the following in a sample of 200 persons randomly selected from each stratum according to the above scheme:

Income group	Sample size	Sample proportion approving policies
upper	10	.60
middle	100	.45
low	90	.20

The overall estimate of the proportion of approval of Reagan's policies is then given by Equation (1.18):

$$p_{st} = \frac{(150 \times .6) + (1500 \times .45) + (1350 \times .2)}{3,000}$$

$$= .345$$

The variance is estimated by:

$$\hat{V}(p_{st}) = \frac{1}{3000^2} \left[\frac{150^2(140)(.6)(4)}{(150)(9)} + \frac{1500^2(1400)(.45)(.55)}{(1500)(99)} \right.$$

$$\left. + \frac{1350^2(1260)(.2)(.8)}{(1350)(89)} \right]$$

$$= \frac{1}{3000^2} [560 + 5250 + 3057.98]$$

$$= .000985$$

The standard deviation is .031.

Example 1.5: To estimate the labor force in a state, the employing units are divided among 4 strata having the numbers:

$$(20, 35, 65, 80)$$

The total cost of a survey includes \$10,000 for administration and data analysis in addition to \$500 per observation in each stratum. The standard deviation of each stratum is known from past experience as:

$$50, \quad 55, \quad 45, \quad 35$$

A sample of 15 units is desired. To find the Neyman allocation in this case, we have

$$N_1 = 20, \quad N_2 = 35, \quad N_3 = 65, \quad N_4 = 80$$

and

$$S_1 = 50, \quad S_2 = 55, \quad S_3 = 45, \quad S_4 = 35$$
$$\sum N_h S_h = (20 \times 50) + (35 \times 55) + (65 \times 45) + (80 \times 35)$$
$$= 8,650$$

Then, $n_1 = (15 \times 20 \times 50)/8,650 = 1.73 \approx 2$, since n_1 has to be an integer. Similarly,

$$n_2 = 3$$
$$n_3 = 5$$
$$n_4 = 5$$

The total cost of this survey is:

$$10,000 + (500 \times 15) = \$17,500$$

Exercises

5. To estimate the total retail sales in a city, a stratified random sample of 30 stores with three strata having (25, 35, 75) stores is to be taken. It is known that there will be a one-time cost of \$50,000, and the costs per observation in different strata will be \$100, \$75, and \$50, respectively. The variances of the retail sales from previous years are known to be 196, 225, and 169, respectively. Find the optimal sample size for each stratum. If the cost for each observation in every stratum happens to be \$200, what will be the optimal allocation? What is the total cost of the survey, using the above allocation schemes?

6. Assuming that Neyman allocation was used in Exercise 5, the following average retail sales were obtained for the three strata, respectively:

$$\$1.25 \text{ million}, \quad \$2.3 \text{ million}, \quad \$3.7 \text{ million}$$

Find the estimate for the total retail sales for the city.

7. Provide the 90% confidence-interval estimates for the total sales given that the sample variances were (in million dollars), respectively, for the strata in Exercises 5 and 6:

$$.4, \quad .6, \quad .5$$

8. The estimate of the proportion of marijuana smokers is to be obtained in a college where the number of students in the following categories (strata) is given with the sample sizes and proportion of smokers:

	Undergraduate		Graduate	
	men	women	men	women
N_h	6,500	5,500	5,000	5,000
n_h	130	110	100	100
p_h	.2	.1	.15	.05

Find the estimate of the proportion of marijuana smokers and its variance in this college.

Ratio and Regression Estimates

Sometimes, we have information on an auxiliary characteristic of a unit and this may improve the estimate of the population characteristic. For example, if we want to estimate the total number of patients with respiratory diseases in a city, we may have auxiliary information about these patients in a previous year, which can be used to improve the estimate.

Let $y_1, y_2, \ldots y_n$ be the simple random sample of n units and let $x_1, x_2, \ldots x_n$ be the corresponding auxiliary information on these units. Then the ratio $\bar{y}/\bar{x}$ can be used to estimate the population ratio R where $R = \dfrac{Y}{X}$. That is,

$$\hat{R} = \frac{\bar{y}}{\bar{x}} \tag{1.23}$$

The estimate of the population total Y when we know the total X is given by:

$$\hat{Y}_R = \frac{\bar{y}}{\bar{x}} X \tag{1.24}$$

The variance of the estimate $\hat{Y}_R$ is given by

$$V(\hat{Y}_R) \approx \frac{N^2(1-f)}{n}(S_y^2 + R^2 S_x^2 - 2RS_{xy}) \tag{1.25}$$

where

$$S_x^2 = \text{population variance of } x$$
$$S_y^2 = \text{population variance of } y$$
$$S_{xy} = \text{population covariance}$$

The estimate of the variance can be obtained by using the estimates of the quantities in (1.25) from the sample. If ratio estimates are obtained in stratified sampling for each stratum, they can be combined to form the population estimates.

The estimate in Equation (1.24) is a biased estimate of Y. Sampling schemes can be given to make this estimate unbiased.

Example 1.6: The number of personal computers owned by the members of a community of 2,700 persons is to be estimated. The community has 5,000 cars registered. It is known that the number of cars (x) owned by a person is highly correlated with the number of personal computers (y). Data were given for 10 randomly chosen persons:

x	1	3	2	1	2	1	0	3	2	2
y	0	1	1	0	1	0	0	1	1	0

We find that $\sum x_i = 17$ and $\sum y_i = 5$. So $\hat{R} = \dfrac{5}{17}$ and the estimate of total number of personal computers is:

$$\hat{Y} = \frac{5}{17} \times 5000 = 1471$$

The estimate of the variance of the total is obtained in terms of the following.

$$s_x^2 = \frac{\sum(x_i - \bar{x})^2}{n-1} = .9$$

$$s_y^2 = \frac{\sum(y_i - \bar{y})^2}{n-1} = .2777778$$

$$s_{xy} = \frac{\sum(x_i - \bar{x})(y_i - \bar{y})}{n-1} = .3888889$$

$$\hat{R} = .2941176$$

$$\hat{V}(\hat{Y}_R) = \frac{2700^2(1 - .0037037)}{10}[.2777778 + (.9 \times .2941176^2)$$
$$- (2 \times .3888889 \times .2941176)]$$
$$= (270)(2700)(.9963963)(.1268742)$$
$$\approx 92149$$

The standard deviation is 304.

Regression Estimates

In place of the ratio estimates, we can obtain the estimates for total of the population with the help of regression if information on an auxiliary variable is available. We shall assume that the population characteristic is linearly related to the auxiliary variable. Further, we assume that the regression coefficient β is known. So we give the regression estimate of the population average by

$$\bar{y}_{lr} = \bar{y} + \beta(\bar{X} - \bar{x}) \qquad (1.26)$$

where $\bar{y}_{lr}$ denotes the *linear regression* estimate and $\bar{X}$ is the known population mean of xs. The estimates of the total are $N\bar{y}_{lr}$.

The variance of the estimate is given by

$$V(\bar{y}_{lr}) = \frac{1-f}{n(N-1)}\sum[y_i - \bar{y} - \beta(\bar{X} - \bar{x})]^2 = \frac{(1-f)}{n}(S_y^2 + \beta_0^2 S_x^2 - 2\beta S_{xy}) \qquad (1.27)$$

where S_y, S_x, and S_{xy} have been defined before.

When the regression coefficient is not known, we estimate it by the least-squares method, discussed in Chapter 9, Volume I. It is given by:

$$\beta_0 = b = \frac{\sum(x_i - \bar{x})(y_i - \bar{y})}{\sum(x_i - \bar{x})^2}$$

In this case, the estimate is:

$$\bar{y}_{lr} = \bar{y} + b(\bar{X} - \bar{x}) \qquad (1.28)$$

The variance of the estimate in (1.28) cannot be easily obtained. However, when n large, the approximate variance is given by

$$V(\bar{y}_{lr}) = \frac{(1-f)}{n}S_y^2(1 - \rho^2) \qquad (1.29)$$

where ρ is the population correlation coefficient between x and y variables and S_y^2 has been given earlier. Its estimate can be obtained in terms of s_y^2 and the sample correlation coefficient. As in the case of the ratio estimate, the regression estimate of Y is also biased. There are sampling schemes other than simple random sampling that make the estimate unbiased. For a recent discussion, see Singh and Srivastava (1980).

Example 1.7: The regression estimate of the total from the data given in Example 1.6 is obtained with the help of the equation:

$$b = \frac{.3888889}{.9} = .4320988$$

$$\bar{x} = 1.7$$

$$\bar{y} = .5$$

$$\bar{X} = 5000/2700 = 1.85$$

$$\bar{y}_{lr} = .5 + .4320988(1.85 - 1.7)$$

$$= .565615$$

Hence, $\hat{Y}_{lr} = 2700 \times .565615 = 1527.$

This obviously differs from the ratio estimate, since in this estimate we have utilized the value of the regression coefficient.

The estimate of variance of $\bar{y}_{lr}$ is given by (approximately):

$$N = 2700, \quad \hat{\rho} = .7777778, \quad f = .0037037$$

And hence,

$$\hat{V}(\hat{Y}) = \frac{N^2(1 - .0037037)}{10}(.2777778)(1 - .7777778^2)$$

$$\approx 79704$$

The standard deviation is 282. The results of regression and ratio estimates are comparable. The regression estimate takes into account the covariance between x and y, which the ratio estimate does not.

Exercises

9. The value of a home in a township is assessed every two years. The data for year 1981 (x) are given. The present assessed value (y) is given for a random sample of 10 homes in the township with 500 homes. For the following data, find the estimate of the total assessed value for the township using ratio and regression estimates. It is known that total X for 1981 is \$4,250 (in thousands). Find the estimate of variance of your estimates.

x	40	60	85	90	45	75	42	35	83	90
y	45	63	80	85	51	78	48	42	85	95

10. In order to estimate the total amount of shipment of household goods by the numbers of Independent Truckers Association in 1983, a random sample of 100 was taken out of a total of 10,500 trucks that do interstate business. The total amount (in million tons) is known for 1982: $X = 250,000.$ We are given that

$$\sum x_i = 2,500, \quad \sum y_i = 2800$$
$$\sum x_i y_i = 75,000$$
$$\sum x_i^2 = 65,100 \quad \sum y_i^2 = 82,200$$

Find the estimate of the total shipments in 1983 and the estimate of its variance using ratio and regression estimates.

11. The number of errors in income tax returns submitted to the Internal Revenue Service for the past year is known to be 2.5 on the average on the basis of 150,000 total returns examined. This year for a given community, out of a total of 1,000 returns submitted, a random sample of 15 returns was made. Data were obtained through comparisons with the same

returns last year:

Last year (x)	2	4	5	6	0	1	2	3	4	3	3	4	0	1	3
This year (y)	3	2	1	0	3	1	2	1	2	2	1	2	5	2	1

Find the estimate of the population average number of errors per return using the regression estimate and obtain the estimate of its variance.

Cluster Sampling

Several methods other than simple random sampling and stratified random sampling are used in practice. Many times the population to be surveyed cannot be easily divided among homogeneous strata and we resort to *cluster sampling*. If a survey is to be done for the whole country, it is convenient to divide it among states, for example, for administrative convenience. The states are then called clusters. If clusters are too large, a further second-stage sampling is needed to obtain information on a smaller subset of items. This is known as *two-stage sampling*. Situations in practice may require several stages for sampling.

Cluster sampling is the method of choosing a group of units and the process of obtaining information on *each* of its units. Generally, a cluster consists of heterogeneous units as compared to the homogeneous units in the stratum under stratified sampling.

When primary units in a cluster are too big to sample completely, *multistage sampling* is used. An advantage of this method is that the total listing of units in the primary units is needed.

Single-Stage Cluster Sampling

In cluster sampling, we choose a random sample of clusters, and we are supposed to have information on each element of the cluster. Let

N = number of clusters in the population

n = sample size

M_i = number of observations in ith cluster with $\sum_{i=1}^{n} M_i = M$,

and $\bar{M} = \dfrac{M}{N}$, the average cluster size, $m = \sum_{i=1}^{n} M_i$

y_i = total of all the observations in the ith cluster

The estimate of the population mean $\bar{Y}$ is:

$$\bar{y} = \sum_{i=1}^{n} y_i/m \tag{1.30}$$

The estimate of the variance of the mean $\bar{y}$ is:

$$\hat{V}(\bar{y}) = \frac{1}{n\bar{M}^2}(1-f)\frac{\sum_{i=1}^{n}(y_i - M_i\bar{y})^2}{n-1} \tag{1.31}$$

The population total will be estimated by $M\bar{y}$ and the estimate of its variance by $M^2\hat{V}(\bar{y})$.

Estimate of Population Proportion

To estimate the proportion p of a characteristic in the sampled clusters, we count the number of units having the specified characteristic. Let

y_i = number of units with the characteristic in the ith cluster

Then we have the estimate of p as

$$\hat{p} = \sum_{i=1}^{n} y_i/m \tag{1.32}$$

and the estimate of the variance of $\hat{p}$ as

$$\hat{V}(\hat{p}) = \frac{1-f}{n\bar{M}^2}\frac{\sum_{i=1}^{n}(y_i - \hat{p}m_i)^2}{n-1} \tag{1.33}$$

Two-Stage Sampling

Since a cluster contains many items, taking a simple random sample can provide enough information to estimate the population total. In two-stage cluster sample, we first obtain a sample of clusters and then we obtain a random sample from each cluster.

In practical situations such as crop-reporting surveys, we have several stages of sampling. For total production estimates of a county, the state forms the first-stage clusters. Since state contains several counties, the second stage is to select some counties and then some farms within a county. One may further want to select a plot of land on the farm in the fourth stage. We consider only two-stage sampling here. Let

$$N = \text{number of clusters in the population}$$
$$n = \text{first-stage sample size}$$
$$M_i = \text{number of elements in cluster } i$$
$$m_i = \text{numbers selected at the second stage}$$
$$y_{ij} = j\text{th observation in the } i\text{th cluster}$$

$$M = \sum_{i=1}^{N} M_i, \quad \bar{M} = M/N, \quad f_i = \frac{m_i}{M_i}$$

$$\bar{y}_i = \sum_{j=1}^{m_i} y_{ij}/m_i, \text{ the average for the } i\text{th cluster}$$

Then the estimate for the population average and estimated variance are given by:

$$\bar{y} = \frac{N}{M} \frac{\sum_{i=1}^{n} M_i \bar{y}_i}{n} \tag{1.34}$$

$$\hat{V}(\bar{y}) = (1 - f)\frac{s_b^2}{n\bar{M}^2} + \frac{1}{n\bar{M}^2 N} \sum_{i=1}^{n} \frac{M_i^2}{m_i}(1 - f_i)s_i^2$$

where

$$s_b^2 = \frac{\sum_{i=1}^{n} (M_i \bar{y}_i - \bar{M}\bar{y})^2}{n - 1} \tag{1.35}$$

and

$$s_i^2 = \frac{\sum_{j=1}^{m_i} (y_{ij} - \bar{y}_i)^2}{m_i - 1}, \quad i = 1, 2, \ldots, n$$

Example 1.8: A market survey is made in a city to estimate the amount of money spent on cosmetics. Since a listing of households in the city is not available, the city is divided in blocks, which are regarded as clusters. Suppose the city is divided into 80 blocks. A sample of 10 blocks is taken and data on cosmetics are collected from each household in the block. The totals are given for each block.

Blocks	1	2	3	4	5	6	7	8	9	10
Number of households (m_i)	50	45	35	42	55	60	40	45	35	50
Total amount spent (y_i)	230	400	275	270	510	525	380	410	401	490

The estimate of the amount spent on cosmetics on the average is given by Equation (1.30).

$$\bar{y} = 3891/457 = 8.51$$

For the estimate of the variance of $\bar{y}$, we need $\bar{M}$. Since we do not know the total number of households in the city, we estimate $\bar{M}$ by:

$$\bar{M} = \sum_{i=1}^{n} m_i/n = 457/10 = 45.7$$

It can be verified that

$$\sum_{i=1}^{n} (y_i - m_i\bar{y})^2 = \sum_{i=1}^{n} y_i^2 - 2\bar{y} \sum_{i=1}^{n} m_i y_i + \bar{y}^2 \sum_{i=1}^{n} m_i^2$$

We have,

$$\sum_{i=1}^{10} y_i^2 = 1,610,551$$

$$\sum_{i=1}^{10} m_i y_i = 182,200$$

$$\sum_{i=1}^{10} m_i^2 = 21,489$$

so that

$$\sum_{i=1}^{10} (y_i - m_i\bar{y})^2 = 65,748.6$$

$$\hat{V}(\bar{y}) = \left(\frac{(80-10)}{80 \times 10 \times (45.7)^2} \right)\left(\frac{65,748.6}{9} \right) = .3061$$

Exercises

12. In Example 1.8, the information of households includes the proportion of rented households. Find the estimate of the rented households in the city given the data:

Block	1	2	3	4	5	6	7	8	9	10
Number of households	50	45	35	42	55	60	40	45	35	50
Proportion of rented households	.2	.1	.3	.4	.1	.2	.5	.2	.1	.5

13. The income of college students and their parents is to be estimated in a state. A cluster sample of 5 colleges out of a total of 350 colleges is taken and a random sample of students from each college is taken to find the

family income. Find the estimate of total income of college students and their families from the two-stage cluster sample.

College	1	2	3	4	5
Total number of students	10,000	12,000	5,000	8,000	25,000
Number of sampled students	25	30	15	12	28
Total income (in thousands)	750	660	810	410	895

14. There are 120 departments in a university. A two-stage cluster sample is taken to estimate the total amount of money spent by the faculty on books in a given year. Data are obtained from a sample of 6 departments. Find the estimate of the total amount spent by the faculty on books and an estimate of its variance.

Total number of faculty	10	15	5	90	20	31
Numbers sampled	2	3	1	6	2	5
Expenses on books	25	75	180	150	105	105
	35	28		100	75	85
		105		80		75
				120		35
				145		110
				80		

Chapter Exercises

1. A simple random sample of 25 students in a high school with a total enrollment of 850 was taken. They were asked if they owned a personal microcomputer. The proportion in the sample was .2 who owned a personal computer. Estimate the total number of personal computers owned by the students. Give a 95% confidence interval for the total.

2. Estimate the average family income in a city where a survey was made by cluster sampling. The city is divided among 50 blocks and a random sample of 10 blocks was taken. The data are given

Block	1	2	3	4	5	6	7	8	9	10
Number of families	10	15	20	35	20	50	60	15	20	17
Total income (in ten thousands)	31.1	22.5	38.1	71.2	41.2	125.6	151.0	30.1	52.7	52.5

3. A survey was made to estimate the total income of persons employed in a state. The labor force is divided into four strata: government employees (including teachers), industrial employees, business employees, and others. It is known that number of employees in the various strata are 129,000, 57,000 180,000, and 25,000. The cost of the survey includes $1,000 in fixed costs and $10 per observation. A sample of 240 units is desired. Given that the variances of the strata are $2,500, $16,900, $14,400, and $22,500, find the sample sizes in each stratum using Neyman allocation.

Summary

Sample survey methods usually are used to estimate the *average, total,* or *proportion* of certain characteristic of a finite population. A *simple random sampling* scheme is used when it is known that the population is homogeneous. When the population can be grouped in various *strata*, a random sample from each stratum is obtained and this is known as *stratified random sampling*. These estimates can be improved if auxiliary information is available on a unit. In that case, additional information can be used to give *ratio* and *regression* estimates. When the population cannot be easily divided into homogeneous strata, *cluster* sampling can be used to obtain information on the whole cluster. If a random sample from a cluster is to be obtained at a second stage, we have *two-stage cluster sampling*. The concept can be extended to multistage cluster sampling.

References

Cochran, W. G. *Sampling Techniques*, 3d. ed. New York: John Wiley & Sons, 1977.
Hansen, Morris H.; Hurwicz, William N.; and Madow, William G. *Sample Survey Methods and Theory*, New York: John Wiley & Sons, 1953.
Kish, Leslie. *Survey Sampling*, New York: John Wiley & Sons, 1965.
Sing, Padam, and Srivastava, A. K. Sampling schemes providing unbiased regression estimates, *Biometrika*, 1980, 67, 205–9.
Sudman, Seymour. *Applied Sampling*, New York: Academic Press, 1976.
Sukhatme, P. V., and Sukhatme, B. V. *Sampling Theory of Surveys with Applications*, Ames: Iowa State University Press, 1970.

chapter two

Life Tables

The mortality and morbidity experience of a population is an important factor in many social and medical applications. To investigate the therapeutic effect of a treatment for a chronic disease, we study the mortality experience of a selected *cohort*. A *cohort* is a group of individuals followed over time for a specific purpose. Such follow-up studies are used in an epidemiologic approach to many problems. Many important associations were discovered as a result of observations on the mortality experience of affected populations. The discovery that cholera was transmitted through water and, more recently, the associations made between lung cancer and cigarette smoking and air pollution and environmental health are results of such mortality studies.

Life tables, developed about four hundred years ago, give the mortality experience of a given population in a standard form. Life-table techniques are being used extensively in medical and health-related studies, such as epidemiology and public health. In this chapter, we consider various aspects of the life table, as well as its construction and uses. Life-table techniques are commonly used in survival analysis see Chapter 3.

Rates

One of the central concepts of life tables is the *crude death rate*. Let P be the number of persons in a community present over a specific period. Let D be the number of deaths during this period. Then by *crude death rate* we mean $\frac{D}{P}k$, where k is an appropriate constant. Usually, $k = 100,000$ for human populations. For example, if a community has 53,280 individuals and there are 382 deaths in the community during 1983, then the crude death rate for 1983 is $(382/53,280) \times 100,000 = 717$ per 100,000 population.

Because of migration, births, and deaths, the total number of persons in a given year is not precisely known. A convention of mortality studies is to use the mid-year population count as the population for the year. We sometimes need death rates, for a specific segment of the population, such as for a certain age group, or a certain disease or occupation, or for a specific sex. The death rates are then obtained for that particular group. In that case, the number of deaths is divided by the *population at risk*. The population *at risk* is the total population of the group we are concerned with. Suppose that during a given year we have D_x deaths in the age group $(x, x + 1)$, where the total number of persons in age group $(x, x + 1)$ is P_x. Then the *age-specific death rate* is given by:

$$q_x = \frac{D_x}{P_x} \times 100,000 \tag{2.1}$$

For example, suppose the number of persons in the age group 45–46 years is 50,000 and there were 139 deaths from heart disease. Then the *disease and age-specific death rate* for 45-year-olds from heart disease is $(139/50,000) \times 100,000$, or 278 per 100,000 persons.

Death rates are used in many applications. The following ranking of 15 major causes of death in the United States for the year 1980 sheds light on the

Table 2.1 Death Rates and Percent of Total Deaths for the 15 Leading Causes of Death: United States 1980 (rates per 100,000 population)

Rank	Cause of death	Rate	Percent of total deaths
	All causes	878.3	100.0
1.	Heart diseases	336.0	38.2
2.	Malignant neoplasms, including neoplasms of lymphatic and hematopoietic tissues	183.9	20.9
3.	Cerebrovascular diseases	75.1	8.6
4.	Accidents and adverse effects	46.7	5.3
5.	Chronic obstructive pulmonary diseases and allied conditions	24.7	2.8
6.	Pneumonia and influenza	24.1	2.7
7.	Diabetes mellitus	15.4	1.8
8.	Chronic liver disease and cirrhosis	13.5	1.5
9.	Atherosclerosis	13.0	1.5
10.	Suicide	11.9	1.4
11.	Homicide and legal intervention	10.7	1.2
12.	Certain conditions originating in the perinatal period	10.1	1.1
13.	Nephritis, nephrotic syndrome, and nephrosis	7.4	0.8
14.	Congenital anomalies	6.2	0.7
15.	Septicemia	4.2	0.5
	All other causes	95.6	10.9

Source: National Center of Health Statistics, U.S. Department of Health, Education & Welfare.

important health problems faced by the nation. The leading cause of death is heart disease, and the second highest is malignant neoplasms (cancer). The complete table for the 15 leading causes of death is given in Table 2.1.

The experience of a cohort, with yearly age-specific death rates, is given in a *complete* life table. When the life table is shortened by taking death rates in five-year intervals, for example, we call it an *abridged life table.*

Complete Life Table

The intervals in the complete life table are one year long. The population at risk in the interval $(x, x + t)$ is denoted by $_tP_x$. Similarly, the number of deaths is given by $_tD_x$ in the interval $(x, x + t)$. Below are all the headings of columns in a standard life table.

(i) *Age interval* $(x, x + t)$. In the complete life table, $t = 1$. However, in the abridged life table, $t = 5$ except for the first year and the next four years. So the intervals for an abridged life table are $(0 - 1)$, $(1 - 5)$, $(5 - 10)$, $(10 - 15)$, and so on.

(ii) *Age-specific death rates,* $_tq_x$. The age-specific death rate is the conditional probability of death of an individual in the age interval $(x, x + t)$, given that an individual has survived to age x, $_tP_x = 1 - _tq_x$ gives the conditional probability of surviving the interval $(x, x + t)$.

(iii) *Number alive at age* x, l_x. This gives the number of persons of age x living at the beginning of the interval $(x, x + t)$. l_0, the number of persons at $x = 0$, is known as the *radix* of the life table. Usually $l_0 = 100,000$ in U.S. life tables. Other values of I_x are calculated based on the initial cohort of 100,000.

(iv) *Number of persons dying in the age interval* $(x, x + t)$, $_td_x$. Note that

$$_td_x = l_x - l_{x+t}$$

and

$$_td_x = l_x {_tq_x} \tag{2.2}$$

(v) *Cumulative number of years lived in the interval* $(x, x + t)$, $_tL_x$. Since people die at different times within the interval $(x, x + t)$, we use half the interval as an average and define

$$_tL_x = l_x - \frac{t}{2} {_td_x}$$

except in the interval $(0, 1)$. In general, we use,

$$_tL_x = \left(l_x - \frac{t}{2} {_td_x} \right) + a_{xt} d_x \tag{2.3}$$

where

$$a_x = \begin{cases} .5, & x > 1 \\ .84, & 0 \le x < 1 \end{cases}$$

(vi) *Number of person years lived beyond age* x, T_x. T_x accumulates the total of all the person years lived beyond age x. That is,

$$T_x = {}_tL_x + {}_tL_{x+t} + {}_tL_{x+2t} + \cdots \qquad (2.4)$$

The values of T_x are computed from ${}_tL_x$ backwards, starting from the last interval of the life table.

(vii) *Life expectancy at age* x, $\mathring{e}_x$. The average remaining time life at age x is called the *expected length of life* at age x. That is,

$$\mathring{e}_x = T_x/l_x \qquad (2.5)$$

Example 2.1: The Ohio life table for the year 1969–71 is given in Table 2.2. The expected number of years to be lived by a person 50 years of age is:

$$\mathring{e}_{50} = 25.59$$

That is, the person is expected to die at age

$$50 + \mathring{e}_{50} = 50 + 25.59 = 75.59$$

The probability that he will die at age 60 is the conditional probability that he dies in the interval 60–61, given that he has survived past age 50:

$$\frac{l_{61} - l_{60}}{l_{50}} = \frac{80,181 - 78,794}{89,739}$$
$$= .0154$$

Survival function: A detailed account of the survival function is given in Chapter 3. Here, the life table allows us to estimate *the survival function*, the proportion of those who survive to age x, given by:

$$\hat{p}_x = \frac{l_x}{l_0}$$

The graph of $\hat{p}_x$ provides an estimate of survival experience for the cohort.

United States Life Tables

Almost all countries have life tables for their populations. Life tables in the United States are prepared under the guidance of the National Center of Health Statistics and the Social Security Administration. The decennial population census provides accurate estimates of the population. However, yearly estimates of populations as well as the number of deaths are available to

Table 2.2 Life Table for the Total Population: Ohio, 1969–71

Age in years	Proportion dying	Of 100,000 born alive		Stationary population		Average remaining lifetime
Period of life between two exact ages stated	Proportion of persons alive at beginning of year of age dying during year	Number living at beginning of year of age	Number dying during year of age	In year of age	In this year of age and all subsequent years	Average number of years of life remaining at beginning of year of age
x to $x+1$	q_x	l_x	d_x	L_x	T_x	$\mathring{e}_x$
0–1	0.01875	100,000	1,875	98,374	7,082,262	70.82
1–2	.00106	98,125	104	98,073	6,983,888	71.17
2–3	.00075	98,021	74	97,984	6,885,815	70.25
3–4	.00062	97,947	61	97,917	6,787,831	69.30
4–5	.00053	97,886	51	97,861	6,689,914	68.34
5–6	.00048	97,835	47	97,811	6,592,053	67.38
6–7	.00045	97,788	44	97,765	6,494,242	66.41
7–8	.00042	97,744	42	97,723	6,396,477	65.44
8–9	.00038	97,702	37	97,684	6,298,754	64.47
9–10	.00034	97,665	33	97,648	6,201,070	63.49
10–11	.00030	97,632	29	97,618	6,103,422	62.51
11–12	.00029	97,603	28	97,589	6,005,804	61.53
12–13	.00032	97,575	32	97,559	5,908,215	60.55
13–14	.00042	97,543	41	97,523	5,810,656	59.57
14–15	.00057	97,502	56	97,474	5,713,133	58.59
15–16	.00075	97,446	73	97,409	5,615,659	57.63
16–17	.00093	97,373	91	97,327	5,518,250	56.67
17–18	.00107	97,282	104	97,230	5,420,923	55.72
18–19	.00117	97,178	113	97,122	5,323,693	54.78
19–20	.00122	97,065	118	97,005	5,226,571	53.85

Age						
20–21	.00126	96,947	123	96,886	5,129,566	52.91
21–22	.00132	96,824	127	96,760	5,032,680	51.98
22–23	.00135	96,697	131	96,631	4,935,920	51.05
23–24	.00136	96,566	131	96,500	4,839,289	50.11
24–25	.00134	96,435	130	96,370	4,742,789	49.18
25–26	.00131	96,305	126	96,242	4,646,419	48.25
26–27	.00128	96,179	123	96,118	4,550,177	47.31
27–28	.00127	96,056	122	95,995	4,454,059	46.37
28–29	.00128	95,934	122	95,873	4,358,064	45.43
29–30	.00131	95,812	126	95,749	4,262,191	44.49
30–31	.00137	95,686	131	95,620	4,166,442	43.54
31–32	.00143	95,555	137	95,487	4,070,822	42.60
32–33	.00151	95,418	143	95,346	3,975,335	41.66
33–34	.00160	95,275	153	95,199	3,879,989	40.72
34–35	.00170	95,122	162	95,041	3,784,790	39.79
35–36	.00183	94,960	174	94,873	3,689,749	38.86
36–37	.00199	94,786	188	94,693	3,594,876	37.93
37–38	.00218	94,598	206	94,494	3,500,183	37.00
38–39	.00241	94,392	228	94,278	3,405,689	36.08
39–40	.00267	94,164	252	94,038	3,311,411	35.17
40–41	.00293	93,912	275	93,774	3,217,373	34.26
41–42	.00321	93,637	301	93,487	3,123,599	33.36
42–43	.00351	93,336	327	93,172	3,030,112	32.46
43–44	.00384	93,009	357	92,831	2,936,940	31.58
44–45	.00421	92,652	390	92,457	2,844,109	30.70
45–46	.00460	92,262	424	92,050	2,751,652	29.82
46–47	.00501	91,838	460	91,608	2,659,602	28.96
47–48	.00547	91,378	501	91,127	2,567,994	28.10
48–49	.00599	90,877	544	90,605	2,476,867	27.26
49–50	.00657	90,333	594	90,036	2,386,262	26.42
50–51	.00722	89,739	648	89,415	2,296,226	25.59
51–52	.00792	89,091	706	88,739	2,206,811	24.77
52–53	.00870	88,385	768	88,001	2,118,072	23.96
53–54	.00953	87,617	835	87,199	2,030,071	23.17
54–55	.01043	86,782	906	86,329	1,942,872	22.39

(Continued)

Table 2.2 (Continued)

Age in years	Proportion dying	Of 100,000 born alive		Stationary population		Average remaining lifetime
Period of life between two exact ages stated	Proportion of persons alive at beginning of year of age dying during year	Number living at beginning of year of age	Number dying during year of age	In year of age	In this year of age and all subsequent years	Average number of years of life remaining at beginning of year of age
x to $x+1$	q_x	l_x	d_x	L_x	T_x	$\mathring{e}_x$
55–56	.01140	85,376	979	85,386	1,856,543	21.62
56–57	.01246	84,397	1,058	84,369	1,771,157	20.86
57–58	.01357	83,339	1,137	83,270	1,686,788	20.12
58–59	.01474	82,702	1,219	82,092	1,603,518	19.39
59–60	.01597	81,483	1,302	80,832	1,521,426	18.67
60–61	.01729	80,181	1,387	79,488	1,440,594	17.97
61–62	.01872	78,794	1,475	78,057	1,361,106	17.27
62–63	.02027	77,319	1,567	76,536	1,283,049	16.59
63–64	.02198	75,752	1,665	74,919	1,206,513	15.93
64–65	.02388	74,087	1,769	73,202	1,131,594	15.27
65–66	.02598	72,318	1,879	71,379	1,058,392	14.64
66–67	.02828	70,439	1,992	69,443	987,013	14.01
67–68	.03075	68,447	2,105	67,395	917,570	13.41
68–69	.03331	66,342	2,210	65,237	850,175	12.82
69–70	.03594	64,132	2,305	62,980	784,938	12.24
70–71	.03861	61,827	2,387	60,634	721,958	11.68
71–72	.04148	59,440	2,466	58,207	661,324	11.13
72–73	.04480	56,974	2,552	55,698	603,117	10.59
73–74	.04876	54,422	2,654	53,095	547,419	10.06
74–75	.05335	51,768	2,761	50,388	494,324	9.55
75–76	.05836	49,007	2,861	47,576	443,936	9.06

76–77	.06358	46,146	2,934	44,680	396,360	8.59
77–78	.06908	43,212	2,985	41,719	351,680	8.14
78–79	.07484	40,227	3,010	38,723	309,961	7.71
79–80	.08102	37,217	3,015	35,709	271,238	7.29
80–81	.08795	34,202	3,008	32,697	235,529	6.89
81–82	.09569	31,194	2,985	29,701	202,832	6.50
82–83	.10393	28,209	2,932	26,743	173,131	6.14
83–84	.11248	25,277	2,843	23,856	146,388	5.79
84–85	.12150	22,434	2,726	21,071	122,532	5.46
85–86	.13166	19,708	2,595	18,410	101,461	5.15
86–87	.14360	17,113	2,457	15,885	83,051	4.85
87–88	.15583	14,656	2,284	13,514	67,166	4.58
88–89	.16734	12,372	2,070	11,337	53,652	4.34
89–90	.17824	10,302	1,836	9,384	42,315	4.11
90–91	.18993	8,466	1,608	7,661	32,931	3.89
91–92	.20345	6,858	1,396	6,160	25,270	3.68
92–93	.21768	5,462	1,189	4,868	19,110	3.50
93–94	.23179	4,273	990	3,778	14,242	3.33
94–95	.24500	3,283	804	2,881	10,464	3.19
95–96	.25745	2,479	639	2,160	7,583	3.06
96–97	.26959	1,840	496	1,592	5,423	2.95
97–98	.28024	1,344	376	1,156	3,831	2.85
98–99	.28977	968	281	827	2,675	2.76
99–100	.29869	687	205	585	1,848	2.69
100–101	.30696	482	148	408	1,263	2.62
101–102	.31461	334	105	281	855	2.56
102–103	.32167	229	74	192	574	2.51
103–104	.32817	155	51	130	382	2.46
104–105	.33414	104	35	87	252	2.41
105–106	.33960	69	23	58	165	2.37
106–107	.34460	46	16	38	107	2.34
107–108	.34917	30	10	25	69	2.30
108–109	.35333	20	7	16	44	2.27
109–110	.35712	13	5	10	28	2.24

prepare the tables. For illustrative purpose, the life table of the State of Ohio is given for the year 1969–71 in Table 2.2. The U.S. life table is given for the year 1980 by sex in Tables 2.3 and 2.4. The mortality experience of males and females can be compared with the help of these tables. For example, the expectation of life at age 30 is given by

$$\mathring{e}_{30} = 42.59 \text{ years for males}$$

and

$$\mathring{e}_{30} = 49.25 \text{ years for females}$$

Insurance companies rely heavily on life tables to determine rates and premiums, and many other agencies, such as state retirement funds and college retirement-equity funds, use life tables extensively.

Exercises

1. From the U.S. life table 1959–61, the following rows are given:

$x - x + t$	$_t q_x$	l_x	$_t d_x$	$_t L_x$	T_x	$\mathring{e}_x$
30–31	.00143	94,905	146	94,836	4,097,944	43.18
40–41	.00300	93,064	279	92,925	3,157,035	33.92
50–51	.00774	88,756	687	88,412	2,244,814	25.29

    (i) What is the expected age of death for a 40-year-old person?
    (ii) What is the expected age of death for a 30-year-old?
    (iii) What is the probability that a 40-year-old person will die at age 50?

2. For the U.S. life table given in Table 2.3, answer the questions in Exercise 1.
3. Complete the following life table for an animal population that has the following age intervals. Use the radix, $l_0 = 1,000$.

$x - x + t$	$_t q_x$
0–1	.03
1–2	.002
2–3	.003
3–4	.005
4–5	.008
5–6	.01
6–7	.02
7–8	.3
8–9	.4
9–10	.5

Table 2.3 United States Life Tables by Sex, 1980

x	q_x	l_x	d_x	L_x	T_x	Male $\overset{\circ}{e}_x$
0	.01404	100000	1404	98757	6984541	69.85
1	.00103	98596	102	98545	6885784	69.84
2	.00075	98495	74	98457	6787238	68.91
3	.00059	98420	58	98391	6688781	67.96
4	.00048	98363	47	98339	6590389	67.00
5	.00042	98315	41	98295	6492050	66.03
6	.00038	98274	37	98255	6393756	65.06
7	.00034	98237	34	98220	6295500	64.08
8	.00030	98203	29	98188	6197280	63.11
9	.00025	98174	25	98161	6099092	62.13
10	.00021	98149	21	98139	6000931	61.14
11	.00021	98128	21	98118	5902792	60.15
12	.00030	98108	29	98093	5804674	59.17
13	.00048	98079	47	98055	5706581	58.18
14	.00073	98032	72	97996	5608526	57.21
15	.00101	97960	99	97911	5510530	56.25
16	.00128	97861	125	97799	5412619	55.31
17	.00150	97736	147	97663	5314820	54.38
18	.00167	97589	163	97508	5217158	53.46
19	.00179	97426	174	97339	5119650	52.55
20	.00190	97252	185	97160	5022310	51.64
21	.00201	97068	195	96970	4925151	50.74
22	.00207	96873	201	96772	4828180	49.84
23	.00209	96672	202	96571	4731408	48.94
24	.00207	96469	200	96369	4634838	48.04
25	.00204	96269	196	96171	4538468	47.14
26	.00201	96073	193	95977	4442297	46.24
27	.00197	95880	189	95786	4346320	45.33
28	.00195	95691	187	95598	4250535	44.42
29	.00194	95504	185	95411	4154937	43.51
30	.00193	95319	184	95227	4059526	42.59
31	.00194	95134	184	95042	3964299	41.67
32	.00196	94950	186	94857	3869257	40.75
33	.00200	94765	189	94670	3774399	39.83
34	.00207	94575	195	94478	3679730	38.91
35	.00215	94380	203	94278	3585252	37.99
36	.00227	94177	213	94070	3490974	37.07
37	.00241	93963	226	93850	3396904	36.15
38	.00259	93737	242	93615	3303054	35.24
39	.00279	93494	261	93364	3209439	34.33
40	.00304	93233	283	93092	3116075	33.42
41	.00332	92950	309	92795	3022984	32.52
42	.00364	92641	337	92473	2930188	31.63
43	.00399	92304	368	92120	2837716	30.74
44	.00438	91936	403	91734	2745596	29.86
45	.00482	91533	441	91312	2653862	28.99
46	.00531	91092	483	90850	2562550	28.13

(Continued)

Table 2.3 (*Continued*)

x	q_x	l_x	d_x	L_x	T_x	Male $\overset{\circ}{e}_x$
47	.00586	90608	531	90343	2471700	27.28
48	.00648	90077	584	89786	2381357	26.44
49	.00717	89494	642	89173	2291571	25.61
50	.00793	88852	704	88500	2202398	24.79
51	.00874	88148	770	87763	2113898	23.98
52	.00959	87377	838	86959	2026136	23.19
53	.01046	86540	905	86087	1939177	22.41
54	.01137	85634	974	85147	1853090	21.64
55	.01239	84660	1049	84136	1767943	20.88
56	.01350	83612	1128	83048	1683807	20.14
57	.01466	82483	1209	81879	1600759	19.41
58	.01586	81274	1289	80630	1518880	18.69
59	.01714	79985	1371	79300	1438250	17.98
60	.01852	78614	1456	77886	1358950	17.29
61	.02007	77158	1548	76384	1281064	16.60
62	.02187	75610	1654	74783·	1204681	15.93
63	.02398	73956	1773	73069	1129898	15.28
64	.02635	72183	1902	71232	1056828	14.64
65	.02895	70281	2035	69264	985596	14.02
66	.03170	68246	2163	67165	916333	13.43
67	.03450	66083	2280	64943	849168	12.85
68	.03730	63803	2380	62613	784225	12.29
69	.04018	61423	2468	60189	721611	11.75
70	.04328	58955	2552	57679	661422	11.22
71	.04669	56403	2633	55087	603743	10.70
72	.05039	53770	2709	52415	548656	10.20
73	.05441	51061	2778	49672	496240	9.72
74	.05877	48283	2837	46861	446569	9.25
75	.06354	45445	2887	44001	399705	8.80
76	.06867	42558	2922	41097	355704	8.36
77	.07405	39635	2935	38168	314607	7.94
78	.07967	36700	2924	35238	276439	7.53
79	.08562	33776	2892	32330	241201	7.14
80	.09199	30884	2841	29464	208870	6.76
81	.09896	28043	2775	26656	179406	6.40
82	.10673	25268	2697	23920	152751	6.05
83	.11539	22571	2605	21269	128831	5.71
84	.12486	19967	2493	18720	107562	5.39
85	.13497	17474	2358	16295	88841	5.08
86	.14558	15115	2200	14015	72547	4.80
87	.15661	12915	2023	11904	58532	4.53
88	.16806	10892	1831	9977	46628	4.28
89	.17997	9062	1631	8246	36651	4.04
90	.19238	7431	1430	6716	28405	3.82
91	.20537	6001	1232	5385	21688	3.61
92	.21898	4769	1044	4247	16303	3.42
93	.23328	3725	869	3290	12057	3.24

(*Continued*)

Table 2.3 (*Continued*)

x	q_x	l_x	d_x	L_x	T_x	Male $\mathring{e}_x$
94	.24829	2856	709	2501	8766	3.07
95	.26305	2147	565	1864	6265	2.92
96	.27742	1582	439	1363	4401	2.78
97	.29121	1143	333	977	3038	2.66
98	.30428	810	247	687	2062	2.54
99	.31645	564	178	475	1375	2.44
100	.32911	385	127	322	900	2.34
101	.34227	259	88	214	578	2.24
102	.35596	170	61	140	364	2.14
103	.37020	110	41	89	224	2.05
104	.38501	69	27	56	135	1.96
105	.40041	42	17	34	79	1.87
106	.41643	25	11	20	45	1.79
107	.43308	15	6	12	25	1.71
108	.45041	8	4	7	14	1.63
109	.46842	5	2	4	7	1.55
110	.48716	2	1	2	4	1.48
111	.50665	1	1	1	2	1.41
112	.52691	1	0	0	1	1.34
113	.54799	0	0	0	0	1.27
114	.56991	0	0	0	0	1.21
115	.59271	0	0	0	0	1.15
116	.61641	0	0	0	0	1.09
117	.64107	0	0	0	0	1.03
118	.66671	0	0	0	0	.97
119	.69338	0	0	0	0	.92

Table 2.4 **United States Life Table by Sex, 1980**

x	q_x	l_x	d_x	L_x	T_x	Female $\mathring{e}_x$
0	.01122	100000	1122	99012	7753142	77.53
1	.00090	98878	89	98834	7654131	77.41
2	.00056	98790	55	98762	7555297	76.48
3	.00042	98735	41	98714	7456535	75.52
4	.00033	98693	33	98677	7357821	74.55
5	.00030	98661	30	98646	7259144	73.58
6	.00028	98631	28	98617	7160498	72.60
7	.00026	98603	26	98590	7061882	71.62
8	.00024	98577	24	98565	6963292	70.64
9	.00022	98553	22	98542	6864727	69.66
10	.00020	98531	20	98521	6766185	68.67
11	.00020	98511	19	98502	6667664	67.68
12	.00021	98492	21	98482	6569162	66.70

(*Continued*)

Table 2.4 (*Continued*)

x	q_x	l_x	d_x	L_x	T_x	Female $\hat{e}_x$
13	.00026	98471	25	98459	6470680	65.71
14	.00032	98446	32	98430	6372221	64.73
15	.00040	98414	39	98395	6273791	63.75
16	.00047	98375	46	98352	6175396	62.77
17	.00053	98329	52	98303	6077045	61.80
18	.00056	98277	55	98249	5978742	60.84
19	.00057	98222	56	98194	5880492	59.87
20	.00058	98166	57	98137	5782298	58.90
21	.00059	98109	58	98080	5684161	57.94
22	.00061	98051	60	98021	5586081	56.97
23	.00062	97991	61	97960	5488061	56.01
24	.00064	97930	62	97899	5390100	55.04
25	.00065	97867	64	97835	5292202	54.08
26	.00067	97803	66	97771	5194366	53.11
27	.00069	97738	67	97704	5096596	52.15
28	.00070	97671	69	97636	4998892	51.18
29	.00072	97602	70	97567	4901256	50.22
30	.00075	97531	73	97495	4803689	49.25
31	.00078	97459	76	97421	4706194	48.29
32	.00082	97383	80	97343	4608773	47.33
33	.00086	97303	84	97261	4511431	46.36
34	.00091	97219	89	97175	4414170	45.40
35	.00098	97130	95	97083	4316995	44.45
36	.00105	97036	102	96985	4219912	43.49
37	.00115	96934	112	96878	4122927	42.53
38	.00128	96822	124	96760	4026049	41.58
39	.00144	96698	139	96628	3929290	40.63
40	.00161	96558	156	96481	3832662	39.09
41	.00180	96403	174	96316	3736181	38.76
42	.00200	96229	192	96133	3639865	37.83
43	.00221	96037	212	95931	3543732	36.90
44	.00242	95825	232	95709	3447802	35.98
45	.00266	95592	254	95465	3352093	35.07
46	.00292	95338	279	95199	3256628	34.16
47	.00320	95059	304	94907	3161429	33.26
48	.00349	94755	331	94590	3066522	32.36
49	.00380	94425	359	94245	2971932	31.47
50	.00413	94066	389	93872	2877687	30.59
51	.00450	93677	422	93466	2783815	29.72
52	.00490	93255	457	93027	2690349	28.85
53	.00533	92798	495	92551	2597322	27.99
54	.00581	92303	536	92035	2504771	27.14
55	.00632	91767	580	91477	2412736	26.29
56	.00689	91187	628	90873	2321258	25.46
57	.00749	90559	678	90220	2230385	24.63
58	.00811	89881	729	89516	2140165	23.81
59	.00878	89152	783	88760	2050649	23.00

(*Continued*)

Table 2.4 (Continued)

x	q_x	l_x	d_x	L_x	T_x	Female $\mathring{e}_x$
60	.00952	88369	841	87948	1961889	22.20
61	.01033	87528	905	87075	1873941	21.41
62	.01124	86623	973	86136	1786865	20.63
63	.01223	85650	1047	85126	1700729	19.86
64	.01332	84602	1127	84039	1615603	19.10
65	.01455	83476	1214	82868	1531564	18.35
66	.01590	82261	1308	81607	1448695	17.61
67	.01730	80954	1400	80253	1367088	16.89
68	.01874	79553	1491	78808	1286835	16.18
69	.02028	78062	1583	77271	1208027	15.48
70	.02203	76479	1685	75637	1130756	14.79
71	.02404	74795	1798	73896	1055119	14.11
72	.02623	72997	1915	72040	981223	13.44
73	.02863	71082	2035	70065	909184	12.79
74	.03128	69047	2160	67967	839119	12.15
75	.03432	66887	2296	65740	771152	11.53
76	.03778	64592	2441	63372	705412	10.92
77	.04166	62151	2589	60857	642041	10.33
78	.04597	59562	2738	58193	581184	9.76
79	.05078	56824	2886	55381	522991	9.20
80	.05615	53938	3029	52424	467610	8.67
81	.06214	50910	3164	49328	415186	8.16
82	.06885	47746	3287	46102	365858	7.66
83	.07631	44459	3393	42762	319755	7.19
84	.08455	41066	3472	39330	276993	6.75
85	.09352	37594	3516	35836	237663	6.32
86	.10323	34078	3518	32319	201827	5.92
87	.11367	30560	3474	28823	169508	5.55
88	.12484	27086	3382	25396	140685	5.19
89	.13677	23705	3242	22084	115289	4.86
90	.14948	20463	3059	18933	93205	4.55
91	.16299	17404	2837	15986	74272	4.27
92	.17730	14567	2583	13276	58287	4.00
93	.19243	11985	2306	10831	45011	3.76
94	.20838	9678	2017	8670	34179	3.53
95	.22428	7662	1718	6802	25509	3.33
96	.23992	5943	1426	5230	18707	3.15
97	.25507	4517	1152	3941	13477	2.98
98	.26950	3365	907	2912	9536	2.83
99	.28298	2458	696	2110	6624	2.69
100	.29713	1763	524	1501	4514	2.56
101	.31198	1239	387	1046	3013	2.43
102	.32758	852	279	713	1967	2.31
103	.34396	573	197	475	1254	2.19
104	.36116	376	136	308	780	2.07
105	.37922	240	91	195	472	1.96
106	.39818	149	59	119	277	1.86

(Continued)

Table 2.4   (Continued)

x	q_x	l_x	d_x	L_x	T_x	Female $\mathring{e}_x$
107	.41809	90	38	71	158	1.76
108	.43899	52	23	41	87	1.66
109	.46094	29	14	23	46	1.57
110	.48399	16	8	12	23	1.48
111	.50665	8	4	6	11	1.41
112	.52691	4	2	3	5	1.34
113	.54799	2	1	1	2	1.27
114	.56991	1	0	1	1	1.21
115	.59271	0	0	0	0	1.15
116	.61641	0	0	0	0	1.09
117	.64107	0	0	0	0	1.03
118	.66671	0	0	0	0	.97
119	.69338	0	0	0	0	.92

Abridged Life Tables

Since complete life tables are large—in the United States, the standard is 110 years—they are usually abridged. The standard table has five-year intervals, except the first two. Since there are too many deaths in the first year of life, the first interval is one year and the second is from ages 1–5. The columns of the abridged life table are the same as the columns of a complete life table. The abridged life tables for the U.S. population are given in Table 2.5 for the year 1980 by race and sex. Abridged life tables are used extensively in special cases for cohorts studied for a specific purpose in clinical or animal studies. They use interval $(x, x + n)$ in place of $(x, x + t)$.

Example 2.2: Consider a person alive at age 40. The expected length of life is given for various cases from Table 2.5:

Category	$\mathring{e}_{40}$
All races	36.7
Male	33.5
Female	39.7
White	37.1
White male	34.0
White female	40.1
All other	33.6
All other male	30.3
All other female	36.8
Black	32.5
Black male	29.1
Black female	35.7

Table 2.5   Abridged Life Tables by Race and Sex: United States, 1980

Age interval	Proportion dying	Of 100,000 born alive		Stationary population		Average remaining lifetime
Period of life between two exact ages stated in years (1) x to $x+n$	Proportion of persons alive at beginning of age interval dying during interval (2) $_nq_x$	Number living at beginning of age interval (3) l_x	Number dying during age interval (4) $_nd_x$	In the age interval (5) $_nL_x$	In this and all subsequent age intervals (6) T_x	Average number of years of life remaining at beginning of age interval (7) $\mathring{e}_x$
All races						
0–1	0.0127	100,000	1,266	98,901	7,371,986	73.7
1–5	.0025	98,734	250	394,355	7,273,085	73.7
5–10	.0015	98,484	150	492,017	6,878,730	69.8
10–15	.0015	98,334	152	491,349	6,386,713	64.9
15–20	.0049	98,182	482	489,817	5,895,364	60.0
20–25	.0066	97,700	648	486,901	5,405,547	55.3
25–30	.0066	97,052	638	483,665	4,918,646	50.7
30–35	.0070	96,414	672	480,463	4,434,981	46.0
35–40	.0091	95,742	875	476,663	3,954,518	41.3
40–45	.0139	94,867	1,321	471,250	3,477,855	36.7
45–50	.0222	93,546	2,079	462,857	3,006,605	32.1
50–55	.0351	91,467	3,209	449,811	2,543,748	27.8
55–60	.0530	88,258	4,676	430,230	2,093,937	23.7
60–65	.0794	83,582	6,638	402,081	1,663,707	19.9
65–70	.1165	76,944	8,965	363,181	1,261,626	16.4
70–75	.1694	67,979	11,517	312,015	898,445	13.2
75–80	.2427	56,462	13,702	248,534	586,430	10.4
80–85	.3554	42,760	15,197	175,192	337,896	7.9
85 and over	1.0000	27,563	27,563	162,704	162,704	5.9
Male						
0–1	0.0140	100,000	1,400	98,787	6,995,933	70.0
1–5	.0029	98,600	283	393,749	6,897,146	70.0

(Continued)

Table 2.5 (Continued)

Age interval	Proportion dying	Of 100,000 born alive		Stationary population		Average remaining lifetime
Period of life between two exact ages stated in years (1) x to $x+n$	Proportion of persons alive at beginning of age interval dying during interval (2) $_nq_x$	Number living at beginning of age interval (3) l_x	Number dying during age interval (4) $_nd_x$	In the age interval (5) $_nL_x$	In this and all subsequent age intervals (6) T_x	Average number of years of life remaining at beginning of age interval (7) $\mathring{e}_x$
5–10	.0018	98,317	173	491,124	6,503,397	66.1
10–15	.0019	98,144	188	490,340	6,012,273	61.3
15–20	.0071	97,956	695	488,224	5,521,933	56.4
20–25	.0101	97,261	986	483,870	5,033,709	51.8
25–30	.0098	96,275	939	478,990	4,549,839	47.3
30–35	.0098	95,336	932	474,430	4,070,849	42.7
35–40	.0122	94,404	1,149	469,323	3,596,419	38.1
40–45	.0180	93,255	1,681	462,351	3,127,096	33.5
45–50	.0288	91,574	2,640	451,697	2,664,745	29.1
50–55	.0462	88,934	4,110	435,061	2,213,048	24.9
55–60	.0707	84,824	5,997	409,935	1,777,987	21.0
60–65	.1061	78,827	8,365	374,082	1,368,052	17.4
65–70	.1571	70,462	11,068	325,406	993,970	14.1
70–75	.2259	59,394	13,420	263,862	668,564	11.3
75–80	.3149	45,974	14,476	193,303	404,702	8.8
80–85	.4354	31,498	13,715	121,742	211,399	6.7
85 and over	1.0000	17,783	17,783	89,657	89,657	5.0
Female						
0–1	0.0113	100,000	1,126	99,021	7,748,490	77.5
1–5	.0022	98,874	215	394,990	7,649,469	77.4
5–10	.0013	98,659	126	492,954	7,254,479	73.5

Age					
10–15	.0011	98,533	113	492,411	6,761,525	68.6
15–20	.0027	98,420	261	491,492	6,269,114	63.7
20–25	.0031	98,159	305	490,045	5,777,622	58.9
25–30	.0034	97,854	334	488,463	5,287,577	54.0
30–35	.0042	97,520	412	486,634	4,799,114	49.2
35–40	.0062	97,108	601	484,140	4,312,480	44.4
40–45	.0100	96,507	963	480,283	3,828,340	39.7
45–50	.0160	95,544	1,524	474,134	3,348,057	35.0
50–55	.0247	94,020	2,322	464,624	2,873,923	30.6
55–60	.0369	91,698	3,381	450,481	2,409,299	26.3
60–65	.0558	88,317	4,931	429,930	1,958,818	22.2
65–70	.0828	83,386	6,902	400,651	1,528,888	18.3
70–75	.1261	76,484	9,643	359,605	1,128,237	14.8
75–80	.1937	66,841	12,950	303,049	768,632	11.5
80–85	.3088	53,891	16,639	228,072	465,583	8.6
85 and over	1.0000	37,252	37,252	237,511	237,511	6.4

White

Age					
0–1	0.0110	100,000	1,105	99,038	7,439,722	74.4
1–5	.0023	98,895	227	395,058	7,340,684	74.2
5–10	.0014	98,668	141	492,963	6,945,626	70.4
10–15	.0015	98,527	147	492,324	6,452,663	65.5
15–20	.0050	98,380	489	490,782	5,960,339	60.6
20–25	.0062	97,891	609	487,937	5,469,557	55.9
25–30	.0058	97,282	567	484,983	4,981,620	51.2
30–35	.0060	96,715	581	482,185	4,496,637	46.5
35–40	.0079	96,134	756	478,914	4,014,452	41.8
40–45	.0121	95,378	1,155	474,216	3,535,538	37.1
45–50	.0199	94,223	1,872	466,754	3,061,322	32.5
50–55	.0323	92,351	2,980	454,804	2,594,568	28.1
55–60	.0498	89,371	4,451	436,367	2,139,764	23.9
60–65	.0760	84,920	6,452	409,249	1,703,397	20.1
65–70	.1132	78,468	8,879	371,074	1,294,148	16.5
70–75	.1662	69,589	11,565	320,047	923,074	13.3
75–80	.2407	58,024	13,965	255,799	603,027	10.4
80–85	.3549	44,059	15,634	180,622	347,228	7.9
85 and over	1.0000	28,425	28,425	166,606	166,606	5.9

(Continued)

Table 2.5 (Continued)

Age interval	Proportion dying	Of 100,000 born alive			Stationary population		Average remaining lifetime
Period of life between two exact ages stated in years (1) x to $x + n$	Proportion of persons alive at beginning of age interval dying during interval (2) $_nq_x$	Number living at beginning of age interval (3) l_x	Number dying during age interval (4) $_nd_x$	In the age interval (5) $_nL_x$	In this and all subsequent age intervals (6) T_x	Average number of years of life remaining at beginning of age interval (7) $\overset{\circ}{e}_x$
White, male							
0–1	0.0123	100,000	1,233	98,928	7,068,892	70.7
1–5	.0026	98,767	259	394,478	6,969,964	70.6
5–10	.0016	98,508	160	492,116	6,575,486	66.8
10–15	.0019	98,348	184	491,369	6,083,370	61.9
15–20	.0072	98,164	703	489,235	5,592,001	57.0
20–25	.0095	97,461	927	484,997	5,102,766	52.4
25–30	.0087	96,534	837	480,529	4,617,769	47.8
30–35	.0084	95,697	800	476,553	4,137,240	43.2
35–40	.0104	94,897	984	472,195	3,660,687	38.6
40–45	.0157	93,913	1,470	466,164	3,188,492	34.0
45–50	.0258	92,443	2,384	456,675	2,722,328	29.4
50–55	.0426	90,059	3,833	441,382	2,265,653	25.2
55–60	.0667	86,226	5,754	417,577	1,824,271	21.2
60–65	.1020	80,472	8,212	382,727	1,406,694	17.5
65–70	.1536	72,260	11,101	334,374	1,023,967	14.2
70–75	.2237	61,159	13,682	272,110	689,593	11.3
75–80	.3145	47,477	14,932	199,743	417,483	8.8
80–85	.4363	32,545	14,199	125,737	217,740	6.7
85 and over	1.0000	18,346	18,346	92,003	92,003	5.0

White, female

Age						
0–1	0.0097	100,000	969	99,156	7,812,718	78.1
1–5	.0020	99,031	193	395,676	7,713,562	77.9
5–10	.0012	98,838	119	493,868	7,317,886	74.0
10–15	.0011	98,719	108	493,350	6,824,018	69.1
15–20	.0027	98,611	265	492,434	6,330,668	64.2
20–25	.0029	98,346	283	491,029	5,838,234	59.4
25–30	.0029	98,063	287	489,619	5,347,205	54.5
30–35	.0036	97,776	356	488,043	4,857,586	49.7
35–40	.0054	97,420	524	485,886	4,369,543	44.9
40–45	.0086	96,896	836	482,541	3,883,657	40.1
45–50	.0141	96,060	1,357	477,123	3,401,116	35.4
50–55	.0225	94,703	2,127	468,520	2,923,993	30.9
55–60	.0342	92,576	3,169	455,395	2,455,473	26.5
60–65	.0527	89,407	4,709	435,930	2,000,078	22.4
65–70	.0792	84,698	6,707	407,755	1,564,148	18.5
70–75	.1220	77,991	9,514	367,579	1,156,393	14.8
75–80	.1907	68,477	13,059	311,089	788,814	11.5
80–85	.3075	55,418	17,042	234,762	477,725	8.6
85 and over	1.0000	38,376	38,376	242,963	242,963	6.3

All other

Age						
0–1	0.0192	100,000	1,924	98,341	6,945,537	69.5
1–5	.0036	98,076	349	391,469	6,847,196	69.8
5–10	.0020	97,727	194	488,102	6,455,727	66.1
10–15	.0018	97,533	173	487,299	5,967,625	61.2
15–20	.0046	97,360	451	485,811	5,480,326	56.3
20–25	.0089	96,909	865	482,488	4,994,515	51.5
25–30	.0109	96,044	1,051	477,646	4,512,027	47.0
30–35	.0130	94,993	1,239	472,005	4,034,381	42.5
35–40	.0175	93,754	1,643	464,872	3,562,376	38.0
40–45	.0256	92,111	2,360	454,931	3,097,504	33.6
45–50	.0385	89,751	3,453	440,501	2,642,573	29.4
50–55	.0565	86,298	4,877	419,783	2,202,072	25.5

(Continued)

Table 2.5 (Continued)

| Age interval | Proportion dying | Of 100,000 born alive | | Stationary population | | Average remaining lifetime |
| | | | | | | |
Period of life between two exact ages stated in years (1) x to $x+n$	Proportion of persons alive at beginning of age interval dying during interval (2) $_nq_x$	Number living at beginning of age interval (3) l_x	Number dying during age interval (4) $_nd_x$	In the age interval (5) $_nL_x$	In this and all subsequent age intervals (6) T_x	Average number of years of life remaining at beginning of age interval (7) $\overset{\circ}{e}_x$
55–60	.0796	81,421	6,485	391,436	1,782,289	21.9
60–65	.1099	74,936	8,233	354,676	1,390,853	18.6
65–70	.1459	66,703	9,729	309,651	1,036,177	15.5
70–75	.1996	56,974	11,373	256,638	726,526	12.8
75–80	.2616	45,601	11,929	196,544	469,888	10.3
80–85	.3609	33,672	12,153	136,875	272,344	8.1
85 and over	1.0000	21,519	21,519	135,469	135,469	6.3
All other, male						
0–1	0.0209	100,000	2,086	98,205	6,526,494	65.3
1–5	.0040	97,914	392	390,729	6,428,289	65.7
5–10	.0024	97,522	229	486,988	6,037,560	61.9
10–15	.0022	97,293	210	486,039	5,550,572	57.1
15–20	.0068	97,083	657	483,998	5,064,533	52.2
20–25	.0138	96,425	1,335	478,966	4,580,535	47.5
25–30	.0164	95,091	1,563	471,574	4,101,569	43.1
30–35	.0194	93,523	1,810	463,265	3,629,995	38.8
35–40	.0250	91,713	2,288	453,123	3,166,730	34.5
40–45	.0344	89,430	3,078	439,773	2,713,607	30.3
45–50	.0515	86,352	4,443	421,092	2,273,834	26.3
50–55	.0761	81,909	6,230	394,543	1,852,742	22.6
55–60	.1054	75,679	7,978	358,993	1,458,199	19.3
60–65	.1434	67,701	9,708	314,703	1,099,206	16.2

All other, male—Con.					
65–70	.1880	57,993	10,904	262,986	784,503	13.5
70–75	.2464	47,089	11,601	206,355	521,517	11.1
75–80	.3178	35,488	11,278	148,224	315,162	8.9
80–85	.4257	24,210	10,306	93,880	166,938	6.9
85 and over	1.0000	13,904	13,904	73,058	73,058	5.3
All other, famale					
0–1	0.0176	100,000	1,757	98,481	7,362,520	73.6
1–5	.0031	98,243	305	392,232	7,264,039	73.9
5–10	.0016	97,938	158	489,250	6,871,807	70.2
10–15	.0014	97,780	136	488,597	6,382,557	65.3
15–20	.0025	97,644	244	487,672	5,893,960	60.4
20–25	.0044	97,400	424	485,994	5,406,288	55.5
25–30	.0061	96,976	590	483,477	4,920,294	50.7
30–35	.0076	96,386	732	480,219	4,436,817	46.0
35–40	.0112	95,654	1,067	475,765	3,956,598	41.4
40–45	.0181	94,587	1,708	468,887	3,480,833	36.8
45–50	.0276	92,879	2,561	458,295	3,011,946	32.4
50–55	.0404	90,318	3,645	442,861	2,553,651	28.3
55–60	.0577	86,673	5,002	421,359	2,110,790	24.4
60–65	.0821	81,671	6,704	392,250	1,689,431	20.7
65–70	.1122	74,967	8,410	354,396	1,297,181	17.3
70–75	.1629	66,557	10,845	306,114	942,785	14.2
75–80	.2208	55,712	12,304	247,578	636,671	11.4
80–85	.3180	43,408	13,806	181,844	389,093	9.0
85 and over	1.0000	29,602	29,602	207,249	207,249	7.0
Black					
0–1	0.0214	100,000	2,145	98,147	6,798,966	68.0
1–5	.0038	97,855	372	390,529	6,700,819	68.5
5–10	.0021	97,483	204	486,853	6,310,290	64.7
10–15	.0018	97,279	179	486,016	5,823,437	59.9
15–20	.0046	97,100	451	484,516	5,337,421	55.0
20–25	.0094	96,649	912	481,086	4,852,905	50.2
25–30	.0121	95,737	1,157	475,854	4,371,819	45.7

(Continued)

Table 2.5 (Continued)

| Age interval | Proportion dying | Of 100,000 born alive | | Stationary population | | Average remaining lifetime |
| | Proportion of persons alive at beginning of age interval dying during interval | Number living at beginning of age interval | Number dying during age interval | In the age interval | In this and all subsequent age intervals | Average number of years of life remaining at beginning of age interval |
Period of life between two exact ages stated in years (1) x to $x+n$	(2) $_nq_x$	(3) l_x	(4) $_nd_x$	(5) $_nL_x$	(6) T_x	(7) $\overset{\circ}{e}_x$
30–35	.0150	94,580	1,422	469,504	3,895,965	41.2
35–40	.0200	93,158	1,864	461,369	3,426,461	36.8
40–45	.0291	91,294	2,655	450,136	2,965,092	32.5
45–50	.0434	88,639	3,848	433,978	2,514,956	28.4
50–55	.0627	84,791	5,319	411,161	2,080,978	24.5
55–60	.0874	79,472	6,948	380,551	1,669,817	21.0
60–65	.1192	72,524	8,648	341,570	1,289,266	17.8
65–70	.1558	63,876	9,949	294,934	947,696	14.8
70–75	.2117	53,927	11,415	241,243	652,762	12.1
75–80	.2745	42,512	11,668	182,691	411,519	9.7
80–85	.3773	30,844	11,637	124,008	228,828	7.4
85 and over	1.0000	19,207	19,207	104,820	104,820	5.5
Black male						
0–1	0.0234	100,000	2,335	97,987	6,366,187	63.7
1–5	.0043	97,665	419	389,668	6,268,200	64.2
5–10	.0025	97,246	242	485,569	5,878,532	60.5
10–15	.0023	97,004	221	484,569	5,392,963	55.6
15–20	.0068	96,783	656	482,506	4,908,394	50.7
20–25	.0147	96,127	1,415	477,300	4,425,888	46.0
25–30	.0183	94,712	1,730	469,268	3,948,588	41.7
30–35	.0226	92,982	2,099	459,833	3,479,320	37.4

35–40	.0289	90,883	2,629	448,122	3,019,487	33.2
40–45	.0397	88,254	3,506	432,856	2,571,365	29.1
45–50	.0584	84,748	4,950	411,834	2,138,509	25.2
50–55	.0847	79,798	6,758	382,683	1,726,675	21.6
55–60	.1160	73,040	8,471	344,578	1,343,992	18.4
60–65	.1559	64,569	10,068	298,103	999,414	15.5
65–70	.2019	54,501	11,003	245,198	701,311	12.9
70–75	.2632	43,498	11,449	188,676	456,113	10.5
75–80	.3351	32,049	10,739	132,368	267,437	8.3
80–85	.4438	21,310	9,457	81,589	135,069	6.3
85 and over	1.0000	11,853	11,853	53,480	53,480	4.5
Black male						
0–1	0.0195	100,000	1,949	98,313	7,228,626	72.3
1–5	.0033	98,051	323	391,420	7,130,313	72.7
5–10	.0017	97,728	164	488,183	6,738,893	69.0
10–15	.0014	97,564	135	487,518	6,250,710	64.1
15–20	.0025	97,429	246	486,591	5,763,192	59.2
20–25	.0046	97,183	447	484,857	5,276,601	54.3
25–30	.0066	96,736	641	482,155	4,791,744	49.5
30–35	.0085	96,095	821	478,556	4,309,589	44.8
35–40	.0125	95,274	1,193	473,573	3,831,033	40.2
40–45	.0202	94,081	1,899	465,900	3,357,460	35.7
45–50	.0309	92,182	2,849	454,109	2,891,560	31.4
50–55	.0446	89,333	3,985	437,103	2,437,451	27.3
55–60	.0634	85,348	5,410	413,741	2,000,348	23.4
60–65	.0891	79,938	7,121	382,564	1,586,607	19.8
65–70	.1197	72,817	8,718	342,880	1,204,043	16.5
70–75	.1729	64,099	11,084	293,205	861,163	13.4
75–80	.2323	53,015	12,315	233,979	567,958	10.7
80–85	.3342	40,700	13,603	168,713	333,979	8.2
85 and over	1.0000	27,097	27,097	165,266	165,266	6.1

Exercises

4. For the U.S. Abridged life table given for the year 1980, answer Exercise 1 for a white male.
5. An abridged life table for the year 1980 is to be made. Give the elements in the first five rows for years 0–1, 1–5, 5–10, 10–15, and 15–20 using the complete life table given in Table 14.3.
6. Using Table 14.6, find the expected age of death except by heart disease for a person who is 50 years old in 1959.
7. Using Table 14.6 what is the conditional probability that a person will not die from a malignant neoplasm at age 70 given that he is alive at 65?
8. Complete the following life table for six age intervals. Use the radix as 100,000.

$x, x + t$	$_tq_x$
0–1	.03
1–5	.001
5–10	.002
10–15	.005
15–20	.002
20 and above	1.0

9. Given the following data for a cohort, obtain the average number of years of life remaining at age 25:

$(x, x + t)$	0–1	1–5	5–10	10–15	15–20	20–25	25–30
$_tq_x$	.05	.003	.001	.003	.007	.008	.009

Competing Risks

There are several competing causes of death. The crude death rates given by the life table do not take into account the large number of risks of death that act on the life of an individual simultaneously. Death certificates usually provide a single cause of death; seldom do they give more than one cause. If a heart-disease patient dies in an automobile accident, his cause of death will be listed as an accident. Similarly, a cancer patient who dies from pneumonia will have his cause of death listed as pneumonia. In order to obtain the real contribution of a given cause of death, an adjustment must be made in the crude death rates. The concept of *net death rates* has been developed to reflect the death experience of a cohort when a given risk of death acts alone. In this case, it is assumed that risks of death from other causes are eliminated.

The theory of competing risks has been developed to provide techniques for net death rates. Competing risks may occur in areas other than in mortality studies. For example, in a military operation where a bomber airplane is used, there are several risks acting on the plane, such as (i) the safe return of the bomber, (ii) the bomber being damaged, and (iii) the bomber being shot down by the enemy. Similarly, a student taking a course may be subject to several competing risks, such as (i) losing interest in the course and failing, (ii) successfully completing the course, and (iii) getting married and dropping the course.

In epidemiological studies, however, the theory of competing risks plays an important role. A recent comprehensive monograph on competing risks is by David and Moeschberger (1978).

Net Death Rate

Consider a cohort in which we are interested only in deaths from cancer and heart diseases. The probability of death from a single cause (say, cancer) when it acts alone, while the risk of death from other causes such as heart diseases has been eliminated, is known as the *net death rate of cancer*.

Similarly, we can define the net death rate of heart diseases. Consider the probabilities:

$$q_1 = \text{probability of death from cancer alone}$$
$$q_2 = \text{probability of death from heart disease alone}$$

These are *net death rates* by definition. Let the *crude death rates* be:

$$Q_1 = \text{crude death rate from cancer}$$
$$Q_2 = \text{crude death rate from heart disease}$$

Assuming that these risks act independently of each other, we can see that if the risk of cancer acts first, $Q_1 = q_1$, the probability of surviving from cancer is $1 - q_1$, and hence $Q_2 = q_2(1 - q_1)$, since death from heart disease can occur only if an individual survives from cancer. So:

$$Q_1 + Q_2 = q_1 + q_2 - q_1 q_2$$

If the risk of heart disease occurs first, we have,

$$Q_2 = q_2$$
$$Q_1 = (1 - q_2)q_1$$

with

$$Q_1 + Q_2 = q_2 + q_1 - q_1 q_2$$

The equation gives the same result as before.

The probability of survival from both is $Q_3 = 1 - Q_1 - Q_2$ or:

$$Q_3 = 1 - (q_1 + q_2 - q_1 q_2) = (1 - q_1)(1 - q_2)$$

Since we do not know which risk acts first, it is not possible to obtain net death rates from crude death rates directly. Therefore, special models are used to overcome this difficulty. The model for *instantaneous risk* or *force of mortality* is given first.

Force of mortality means the conditional probability of death at an instant given the individual has survived to that instant. Let $F_X(x)$ be the cumulative probability distribution function of X, the age at death, and let $f_X(x)$ be the probability density function of X. Consider the time interval $(x, x + \Delta x)$. The probability of death in this interval is $f_X(x)\Delta x$. The probability of survival to age x is given by $1 - F_X(x)$. Hence the conditional probability of death in the interval $(x, x + \Delta x)$ given an individual survives to x, is:

$$\frac{f_X(x)\Delta x}{1 - F_X(x)}$$

The *force of mortality* is given by

$$\lambda(x) = \frac{f_X(x)}{1 - F_X(x)} \tag{2.6}$$

In engineering applications, $\lambda(x)$ is called "failure rate" or "hazard rate", and it is used extensively in reliability studies. When the force of mortality does not depend on age x, that is, when $\lambda(x) = \lambda$, the probability density function of X turns out to be exponential. So:

$$f_X(x) = \begin{cases} \lambda e^{-\lambda x}, & x > 0, \\ 0, & \text{elsewhere} \end{cases}$$

We have also seen in Chapter 5, Volume I, that

$$F_X(x) = \begin{cases} 1 - e^{-\lambda x}, & x > 0 \\ 0, & x \le 0 \end{cases}$$

$F_X(x)$ gives the probability of death before age x, so the probability of survival to x is:

$$q = 1 - e^{-\lambda x} \tag{2.7}$$

The probability of survival after age x is $1 - q_x = e^{-\lambda x}$. Note that equation (2.6) can be easily verified in the above case, since $f_X(x)/[1 - F(x)] = \lambda e^{-\lambda x}/e^{-\lambda x} = \lambda$ as required. Suppose now that $\lambda = \lambda_1$ for cancer and $\lambda = \lambda_2$ for heart disease. Then for one-year age intervals, with $x = 1$, we have:

$$q_1 = 1 - e^{-\lambda_1}$$
$$q_2 = 1 - e^{-\lambda_2}$$

The probability of survival from cancer and heart diseases, then, is:

$$(1 - q_1)(1 - q_2) = e^{-\lambda_1} e^{-\lambda_2} = e^{-(\lambda_1 + \lambda_2)} \tag{2.8}$$

If there are more than two causes, say k causes of death with forces of mortality $\lambda_1, \lambda_2, \ldots \lambda_k$, the probability of survival is:

$$e^{-(\lambda_1 + \lambda_2 + \cdots + \lambda_k)}$$

That is, the "overall" force of mortality is $\lambda_1 + \lambda_2 + \cdots + \lambda_k$. Such a model of competing risks is called *additive*.

Estimates for Force of Mortality

In Chapter 5, Volume I, we have seen that the mean of an exponential distribution with parameter λ is $\frac{1}{\lambda}$. Given a random sample from the exponential population, the maximum likelihood estimate of the parameter λ is the reciprocal of the sample average.

Let t_{ij} be the time of death of an individual j from cause i. Let n_i be the number of individuals dying from cause i. Then an estimate of λ_i is given by:

$$\lambda_i = \frac{n_i}{\sum\limits_{i=1}^{n_i} t_{ij}}, \quad i = 1, 2, \ldots, k$$

A large sample approximation can be used to obtain the confidence intervals for λ_i.

The $(1 - \alpha)$-level confidence interval for λ_i is given by:

$$\left(\hat{\lambda}_i - z_{1-\alpha/2} \sqrt{\frac{\hat{\lambda}_i}{n_i \hat{\lambda}}}, \; \hat{\lambda}_i + z_{1-\alpha/2} \sqrt{\frac{\hat{\lambda}_i}{n_i \hat{\lambda}}} \right) \tag{2.9}$$

where

$$\hat{\lambda} = \sum_{i=1}^{k} \hat{\lambda}_i$$

For the derivation of the above result, see Gross and Clark (1975).

Example 2.3: Suppose the times of death from heart disease and auto accidents are exponentially distributed (with a constant risk of mortality). Of the 50 patients exposed to the risks, 10 have heart disease. They die with an average survival time of 15.2 months from birth. Another 40 patients are exposed to the risk of auto accidents with the average time of death being 40.3 months. We have:

$$\hat{\lambda}_1 = \text{the estimated risk of mortality from heart disease}$$

$$= \frac{1}{15.2} = .0658$$

$$\hat{\lambda}_2 = \text{the estimate of risk of mortality from auto accidents}$$

$$= \frac{1}{40.3} = .0248$$

Table 2.6 Abridged Life Tables Eliminating Specified Causes of Death for the Total Population: United States, 1959–1961

x to $x+n$	Infective and parasitic diseases				Tuberculosis			
	$_nq_x$	t_x	$_nL_x$	$\mathring{e}_x$	$_nq_x$	t_x	$_nL_x$	$\mathring{e}_x$
0–1	0.02559	100,000	97,833	70.11	0.02592	100,000	97,815	69.99
1–5	.00390	97,441	388,855	70.95	.00417	97,408	388,659	70.85
5–10	.00228	97,061	484,710	67.22	.00240	97,002	484,380	67.14
10–15	.00214	96,840	483,733	62.37	.00220	96,769	483,364	62.29
15–20	.00447	96,633	482,173	57.49	.00454	96,556	481,773	57.42
20–25	.00606	96,201	479,571	52.74	.00615	96,118	479,136	52.67
25–30	.00623	95,618	476,619	48.05	.00632	95,527	476,144	47.98
30–35	.00777	95,022	473,351	43.33	.00787	94,923	472,833	43.27
35–40	.01115	94,284	468,968	38.65	.01126	94,176	468,405	38.60
40–45	.01769	93,233	462,338	34.06	.01783	93,116	461,726	34.00
45–50	.02811	91,584	451,954	29.62	.02832	91,455	451,275	29.57
50–55	.04475	89,009	435,678	25.40	.04505	88,865	434,912	25.36
55–60	.06553	85,026	411,975	21.47	.06598	84,862	411,089	21.43
60–65	.09885	79,454	378,592	17.79	.09940	79,263	377,573	17.76
65–70	.14306	71,600	333,380	14.45	.14371	71,384	332,255	14.43
70–75	.20671	61,357	276,016	11.43	.20745	61,125	274,853	11.41
75–80	.30117	48,674	207,372	8.74	.30190	48,445	206,297	8.72
80–85	.44602	34,015	131,402	6.41	.44676	33,819	130,573	6.40
85 and over	1.00000	18,843	86,489	4.59	1.00000	18,710	85,762	4.58
	Malignant neoplasms				Malignant neoplasms of digestive organs			
	$_nq_x$	t_x	$_nL_x$	$\mathring{e}_x$	$_nq_x$	t_x	$_nL_x$	$\mathring{e}_x$
0–1	.02586	100,00	97,818	72.16	.02592	100,000	97,815	70.55
1–5	.00378	97,414	388,752	73.07	.00419	97,408	388,654	71.43
5–10	.00202	97,046	484,704	69.34	.00240	97,000	484,371	67.72
10–15	.00190	96,850	483,842	64.48	.00220	96,767	483,353	62.88
15–20	.00417	96,666	482,412	59.60	.00454	96,554	481,763	58.01

Age		Heart diseases						
20–25	.00573	96,263	479,961	54.84	.00615	96,116	479,125	53.26
25–30	.00569	95,712	477,208	50.14	.00632	95,524	476,131	48.58
30–35	.00681	95,168	474,280	45.41	.00783	94,920	472,826	43.87
35–40	.00934	94,519	470,523	40.70	.01108	94,177	468,443	39.19
40–45	.01431	93,636	465,058	36.06	.01731	93,133	461,914	34.60
45–50	.02212	92,296	456,740	31.55	.02711	91,521	451,842	30.17
50–55	.03485	90,255	443,882	27.20	.04266	89,040	436,254	25.93
55–60	.05117	87,109	425,063	23.09	.06199	85,241	413,719	21.97
60–65	.07848	82,652	397,940	19.19	.09295	79,957	382,117	18.25
65–70	.11682	76,165	359,654	15.60	.13462	72,525	339,200	14.85
70–75	.17534	67,268	308,077	12.31	.19552	62,762	284,134	11.75
75–80	.26597	55,473	241,641	9.38	.28756	50,491	216,968	8.98
80–85	.40981	40,719	161,657	6.84	.43145	35,972	140,517	6.58
85 and over	1.00000	24,032	117,015	4.87	1.00000	20,452	96,139	4.70

Age		Influenza and pneumonia						
0–1	.02586	100,000	97,819	75.78	.02367	100,000	97,919	70.42
1–5	.00415	97,414	388,687	76.79	.00361	97,633	389,705	71.12
5–10	.00236	97,010	484,431	73.10	.00227	97,280	485,804	67.37
10–15	.00214	96,781	483,438	68.27	.00211	97,059	484,834	62.52
15–20	.00441	96,574	481,895	63.41	.00443	96,854	483,287	57.65
20–25	.00592	96,149	479,344	58.68	.00605	96,425	480,694	52.89
25–30	.00593	95,580	476,494	54.01	.00624	95,842	477,735	48.20
30–35	.00697	95,014	473,470	49.32	.00779	95,244	474,449	43.49
35–40	.00913	94,351	469,712	44.65	.01113	94,502	470,054	38.31
40–45	.01305	93,490	464,576	40.04	.01766	93,450	463,420	34.21
45–50	.01908	92,270	457,221	35.53	.02804	91,800	453,034	29.78
50–55	.02842	90,509	446,457	31.17	.04463	89,225	436,763	25.56
55–60	.03971	87,937	431,409	27.00	.06531	85,243	413,066	21.63
60–65	.05774	84,445	410,650	23.01	.09813	79,675	379,774	17.96
65–70	.08219	79,570	382,240	19.26	.14149	71,857	334,833	14.63
70–75	.11814	73,030	344,520	15.75	.20347	61,689	277,982	11.61
75–80	.17473	64,402	295,092	12.51	.29486	49,137	210,111	8.92
80–85	.26915	53,149	230,657	9.61	.43483	34,649	134,899	6.59
85 and over	1.00000	38,844	280,067	7.21	1.00000	19,582	93,473	4.77

(Continued)

Table 2.6 (Continued)

x to x + n	$_nq_x$	l_x	$_nL_x$	$\mathring{e}_x$	$_nq_x$	l_x	$_nL_x$	$\mathring{e}_x$
	Malignant neoplasms of respiratory system				Diabetes			
0–1	0.02592	100,000	97,815	70.21	0.02592	100,000	97,815	70.11
1–5	.00420	97,408	388,650	71.08	.00419	97,408	388,652	70.97
5–10	.00240	96,999	484,364	67.37	.00239	97,000	484,370	67.26
10–15	.00221	96,765	483,345	62.53	.00219	96,767	483,360	62.41
15–20	.00455	96,552	481,752	57.66	.00452	96,556	481,777	57.55
20–25	.00618	96,113	479,106	52.91	.00613	96,119	479,147	52.79
25–30	.00638	95,519	476,092	48.22	.00630	95,530	476,163	48.10
30–35	.00793	94,910	472,751	43.52	.00787	94,928	472,857	43.39
35–40	.01126	94,157	468,306	38.85	.01129	94,180	468,422	38.72
40–45	.01760	93,097	461,675	34.26	.01786	93,117	461,726	34.13
45–50	.02762	91,459	451,430	29.82	.02829	91,454	451,270	29.70
50–55	.04353	88,933	435,549	25.59	.04483	88,866	434,955	25.49
55–60	.06340	85,062	412,577	21.64	.06534	84,882	411,304	21.56
60–65	.09578	79,669	380,222	17.92	.09797	79,336	378,182	17.88
65–70	.13964	72,038	336,063	14.54	.14134	71,563	333,500	14.54
70–75	.20373	61,978	279,329	11.48	.20401	61,449	276,845	11.51
75–80	.29926	49,352	210,548	8.76	.29765	48,913	208,855	8.80
80–85	.44507	34,582	133,710	6.41	.44259	34,354	133,072	6.44
85 and over	1.00000	19,191	88,124	4.59	1.00000	19,149	88,280	4.61
	Major cardiovascular-renal diseases				Vascular lesions affecting central nervous system			
0–1	.02581	100,000	97,822	80.79	.02589	100,000	97,817	71.18
1–5	.00410	97,419	388,719	81.93	.00416	97,411	388,672	72.06
5–10	.00231	97,020	484,493	78.26	.00238	97,006	484,405	68.36
10–15	.00205	96,796	483,532	73.43	.00217	96,775	483,401	63.52
15–20	.00424	96,598	482,047	68.58	.00449	96,565	481,830	58.65
20–25	.00563	96,188	479,604	63.86	.00608	96,132	479,223	53.90

25–30	.00553	95,646	476,915	59.20	.00624	95,547	476,264	49.21
30–35	.00636	95,118	474,126	54.52	.00771	94,951	473,005	44.51
35–40	.00819	94,513	470,723	49.85	.01096	94,218	468,678	39.83
40–45	.01142	93,740	466,170	45.24	.01715	93,186	462,215	35.24
45–50	.01642	92,669	459,768	40.73	.02698	91,588	452,205	30.81
50–55	.02389	91,148	450,569	36.37	.04253	89,117	436,663	26.59
55–60	.03249	88,970	437,961	32.20	.06167	85,327	414,189	22.66
60–65	.04498	86,080	421,142	28.19	.09142	80,065	382,881	18.97
65–70	.06043	82,208	399,074	24.39	.12970	72,745	341,002	15.62
70–75	.07981	77,240	371,300	20.80	.18289	63,310	288,439	12.56
75–80	.10680	71,075	337,089	17.38	.26059	51,732	225,688	9.79
80–85	.14946	63,485	294,090	14.14	.38392	38,251	154,288	7.35
85 and over	1.00000	53,996	603,802	11.18	1.00000	23,566	126,714	5.38
	Congenital malformations				Motor vehicle accidents			
0–1	.02232	100,000	98,105	70.25	.02585	100,000	97.818	70.44
1–5	.00371	97,768	390,213	70.85	.00381	97,415	388,754	71.34
5–10	.00220	97,405	486,448	67.10	.00199	97,043	484,698	67.58
10–15	.00205	97,191	485,509	62.25	.00185	96,851	483,829	62.70
15–20	.00440	96,991	483,981	57.37	.00288	96,672	482,707	57.82
20–25	.00607	96,564	481,386	52.61	.00405	96,393	481,025	52.98
25–30	.00631	95,979	478,401	47.92	.00503	96,002	478,852	48.18
30–35	.00792	95,373	475,064	43.20	.00695	95,519	476,033	43.41
35–40	.01137	94,618	470,578	38.53	.01051	94,856	471,968	38.70
40–45	.01800	93,542	463,803	33.94	.01716	93,859	465,570	34.08
45–50	.02857	91,858	453,211	29.51	.02768	92,248	455,340	29.63
50–55	.04544	89,234	436,633	25.30	.04451	89,695	439,099	25.39
55–60	.06649	85,179	412,519	21.38	.06552	85,702	415,264	21.45
60–65	.10004	79,515	378,649	17.72	.09899	80,087	381,587	17.77
65–70	.14451	71,561	332,931	14.39	.14336	72,159	335,931	14.44
70–75	.20838	61,219	275,126	11.39	.20711	61,814	278,008	11.42
75–80	.30288	48,462	206,241	8.71	.30154	49,012	208,764	8.73
80–85	.44768	33,784	130,344	6.39	.44652	34,233	132,200	6.49
85 and over	1.00000	18,660	85,404	4.58	1.00000	18,947	86,844	4.58

Source: National Center of Health Statistics.

$$\hat{q}_1 = 1 - e^{-.0658} = .0637$$
$$\hat{q}_2 = 1 - e^{-.0248} = .0345$$

In the United States, abridged life tables are obtained for various population groups when specified causes of death are eliminated. That is, the survival experience is provided only when the specified cause acted alone. Table 2.6 provides abridged life tables when a dozen specified causes are eliminated for the total U.S. population for the year 1959–61. Life expectancies and net probabilities of death from these causes are given by age groups. For example, the net probability of death from tuberculosis during age 45–50 is .02832, where as from heart diseases for the same age group, it is .01908.

Exercises

10. Suppose in a group of patients, 20 patients die from cancer with an average survival time of 25.2 months, and 80 patients die from heart disease with an average survival time of 35.7 months. Both groups are simultaneously exposed to these risks. Find the observed forces of mortality and the net probability of death from cancer and heart disease. Find the 95% confidence interval for the forces of mortality for cancer and heart disease.

11. Two diseases, AIDS and syphilis, are competing for the death of 35 patients in a hospital; 9 die from AIDS. Their survival time are:

$$15, \quad 10, \quad 15, \quad 24, \quad 30, \quad 18, \quad 17, \quad 19, \quad 36$$

The remaining patients died of syphilis with an average survival time of 35.7 months. Find the estimates of their forces of mortality. What is the 90% confidence interval for the force of mortality for AIDS?

Chapter Exercises

1. Complete the following life table using the radix equal to 10,000.

$(x \text{ to } x + t)$	$_tq_x$
0–1	.2
1–5	.001
5–10	.003
10–15	.01
15–20	.2
20 and above	1.0

What is the life expectancy in years for an individual in this population who is 5?

2. Construct a life table for *Drosophila Melanogaster* given the following data for males:

Age intervals (days) x to $x+n$	Number living at x	Number dying in $(x, x+n)$
0–5	270	2
5–10	268	4
10–15	264	3
15–20	261	7
20–25	254	3
25–30	251	3
30–35	248	16
35–40	233	66
40–45	166	36
45–50	130	54
50–55	76	42
55–60	34	21
60 and over	13	13

Source: Chiang (1968).

3. The times of death (in days) of 43 patients from granulocytic leukemia are given below. Find the risk of mortality assuming an exponential model and obtain a 95% confidence interval for it.

7	47	58	74	177	233	273
285	317	429	440	445	455	468
495	497	532	571	579	581	650
702	715	779	881	900	930	968
1,077	1,109	1,314	1,334	1,367	1,534	1,712
1,784	1,877	1,886	2,045	2,056	2,260	2,429
2,509						

Source: Bryson, M. C., and Siddiqui, M. M. (1969), Some Criteria for Aging, *J. Amer. Statist. Assoc.*, 1969, 1472–83.

Summary

Life tables are used to study the mortality experience of a *cohort*. Age-specific death rates utilize the concept of *population at risk* and the number of deaths. A *complete life table* is presented in one-year intervals and has seven columns, headed by *age interval, age-specific death rate, numbers alive at age x, number of persons dying in the age interval, number of years lived in the interval, number of years lived beyond age* x, and *life expectancy at age* x. *Abridged life tables* have the same number of columns, except five-year intervals are used, the first two being 0–1 and 1–5 years. The theory of *competing risks* gives *net rates* when several causes of death are competing for the life of an individual. The *force of mortality*, or *hazard rate*, is an important concept in parametric models, providing the conditional probability of death in an interval given survival to age x. When the force of mortality is constant, then the times of death have an exponential probability distribution. The follow-up studies required in clinical experiments using life tables are given in Chapter 3.

References

Berkson, J., and Elveback, L. Competing exponential risks, with particular reference to smoking and lung cancer, *J. Am. Statist. Assoc.*, 1960, *55*, 415–28.

Chiang, C. L. *Introduction to Stochastic Processes in Biostatistics*. New York: John Wiley & Sons, 1968.

Cornfield, J. A statistical problem arising from retrospective studies, *Third Berkeley Symposium on Mathematical Statistics and Probability*, J. Neyman, ed. Berkeley; University of California Press, 1956, 135–48.

Cutler, S. J., and Ederer, F. Maximum utilization of life table method in analyzing survival, *J. of Chronic Diseases*, 1958, *8*, 699–712.

David, H. A., and Moeschberger, M. L. *The Theory of Competing Risks*. New York: MacMillan, 1978.

Elandt-Johnson, Regina C., and Johnson, Norman L. *Survival Models and Data Analysis*. New York: John Wiley & Sons, 1980.

Fix, E., and Neyman, J. A simple stochastic model of recovery, relapse, death and loss of patients, *Human Biol.*, 1951, *23*, 205–41.

Gehan, Edmund A. Estimating survival functions from the life table, *J. of Chronic Diseases*, 1969, *21*, 629–44.

Gross, Alan J., and Clark, Virginia A. *Survival Distributions: Reliability Applications in the Biomedical Sciences*. New York: John Wiley & Sons, 1975.

Kuzma, J. A comparison of two life table methods, *Biometrics*, 1967, *23*, 51–64.

MacMahon, Brian; Pugh, Thomas F.; and Ipsen, Johannes. *Epidemiologic Method*. Boston: Little, Brown, 1960.

Miller, Ruppert, G., Jr.; Efron, B.; Brown, B. W., Jr.; and Moses, Linconn E. *Biostatistics Casebook*. New York: John Wiley & Sons, 1980.

Spiegelman, Mortimer. *Introduction to Demography*. Cambridge, Mass.: Harvard University Press, 1968.

chapter three

Survival Analysis

In Chapter 2, we discussed complete and abridged life tables as they are used in mortality studies. Life tables provide the life expectancies at a given age for a specific cohort, and this quantity can be used for comparison of the life experience of two or more cohorts. In engineering and biomedical sciences, data on survival or failure times may not be available on all individuals. In follow-up studies, data may be censored for various reasons. The individual may be lost to follow-up or the study may be discontinued before all the responses are available. In some engineering applications, tests for studying failure time of individual items such as light bulbs may be stopped when a prespecified number of them have failed. Methods of dealing with censored data in survival analysis are needed.

A clinical life table contains the mortality experience of a group of patients who enter the study at different times and who leave the study because of death or loss to follow-up. As a rule, the number of observations in such follow-up studies is not large. For a comparison of survival experience under two experimental conditions, we use survival functions. This chapter presents parametric and nonparametric methods to deal with survival functions.

The concepts in reliability theory and survival analysis are similar. The failure rate is used in reliability studies, whereas the survival rate is used in survival analysis. Later in this chapter, these concepts will be described in terms of survival of human populations, although their corresponding meaning in reliability concepts is easy to understand.

Excellent books on survival analysis have appeared recently; for example, see Elandt-Johnson and Johnson (1980), Gross and Clark (1975), Kalbfleish and Prentice (1980), and Miller (1980).

Prevalence and Incidence Rates

Establishing a relationship between various diseases and causes of death is important in epidemiology. The early development of epidemiology was concerned with the problems of epidemics, their infection rates, and problems connected with their spread. However, these methods can now be employed in other areas. The application of the epidemiologic method resulted in many discoveries of relationships, such as between smoking and lung cancer, between air pollution levels and chronic diseases, and between a lack of fluorides in water and dental caries among children. Epidemiologists use several rates described in this chapter. An individual suffering from a disease is generally known as a *case* in epidemiology.

Prevalence rate of a disease. This rate is obtained by the ratio of all the existing cases of the disease at a specific point or for a specific period and the number of individuals at risk in a given community. That is, at a given point or for a period of time,

$$\text{Prevalence rate} = \frac{\text{Number of cases of disease present}}{\text{Total number at risk}} \qquad (3.1)$$

Incidence rate of a disease. This rate is obtained for a period of time. Suppose we are interested in the T-year incidence rate of disease A. Then for the T-year period, we have:

$$\text{Incidence rate} = \frac{\text{Number of cases during } T \text{ years}}{T \times (\text{Average number of persons in a year})} \qquad (3.2)$$

Note that in incidence rate we consider the new cases in the T-year period, while in prevalence rate we use the existing cases at a given point.

Sometimes the same person can have the disease several times in a year, so it is plausible that the numerator can be larger than the denominator. That is, the incidence rate may be greater than one. Usually these rates are expressed per 100 or 1,000 of populations.

Suppose, for example, in a community of 3,697 persons, 5 cases of hepatitis were found in a given year. The *prevalence rate* of hepatitis then is, $\frac{5 \times 1,000}{3,697} = 1.35$ cases per 1,000 of population. Now, suppose the population and number of new cases of hepatitis over a five-year period is:

Year	1979	1980	1981	1982	1983
Population	3,697	3,590	3,825	3,775	3,872
Cases	5	3	0	1	4

The *incidence rate* of hepatitis in this community for a 5-year period per 1,000 of population per year is:

$$\frac{(5 + 3 + 0 + 1 + 4) \times 1,000}{3,697 + 3,590 + 3,825 + 3,757 + 3,872} = \frac{13,000}{18,741} = .69$$

Exponential Model for Survival Functions

Let X be the survival time of a patient. The cumulative distribution function, $F(x)$ of X gives the probability that the event of death occurs before x or:

$$P(X \leq x) = F(x)$$

The probability that the patient survives time x is:

$$S(x) = 1 - F(x)$$

$S(x)$ is called the *survival function* of X. Remember from Chapter 2 that the *force of mortality* or *hazard rate* gives the conditional probability of death is an interval given survival to that interval. If the force of mortality is constant—that is, it does not depend on x—then the probability distribution of X is exponential:

$$f(x) = \begin{cases} \lambda e^{-\lambda x}, & x > 0 \\ 0, & \text{otherwise} \end{cases}$$

Recall that the mean of X is given by:

$$E(x) = \frac{1}{\lambda}$$

The variance of X is given by:

$$\text{Var}(x) = \frac{1}{\lambda^2}$$

In this case of exponential density, we have:

$$F(x) = 1 - e^{-\lambda x}$$

and

$$S(x) = e^{-\lambda x} \tag{3.3}$$

Given a random sample of size n of times of death as $X_1, X_2, \ldots X_n$ for n patients under the above model, the maximum likelihood estimate of λ is:

$$\hat{\lambda} = \frac{1}{\overline{X}} \tag{3.4}$$

The survival function is estimated by:

$$\hat{S}(x) = e^{-\hat{\lambda} x} \tag{3.5}$$

Example 3.1: Five electron tubes are tested to measure their failure times. We assume that failure times are exponentially distributed. The times of failure (in hours) are 1,500, 1,701, 1,389, 503, and 1,407:

$$\overline{X} = 1300.$$

Hence the survival function is:

$$\hat{S}(x) = e^{-x/1300}$$

The probability that a randomly chosen electron tube will last for more than 1,600 hours from Equation (3.3) is:

$$e^{-1600/1300} = e^{-1.23} = .292$$

The probability that a randomly chosen tube will fail before 1,000 hours is:

$$1 - \hat{S}(1000) = .537$$

Often, survival data are censored. Before considering the estimation of parameters of the exponential distribution under censoring, three kinds of censoring mechanisms will be described.

Censoring Mechanisms

Three of the several kinds of censoring mechanisms are type I and II censoring and random censoring.

(i) *Type I censoring*: In Type I censoring, the experiment is terminated at a fixed time. That is, in place of observing the lifetime $X_1, X_2, \ldots X_n$ of n items being tested, we observe only those Xs that are smaller than T, the censoring time. The number of uncensored observations, n_u, is random and $n - n_u$ observations are all bigger than T.

(ii) *Type II censoring*: In Type II censoring, the experiment is terminated when a preassigned number of failures has occurred. Let $X_{(1)} \le X_{(2)} \le \cdots \le X_{(r)}$ be the lifetimes of preassigned number r of test terms, then $(n - r)$

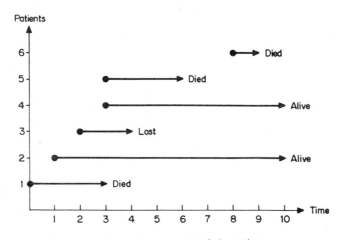

Figure 3.1.   Clinical trial of six patients.

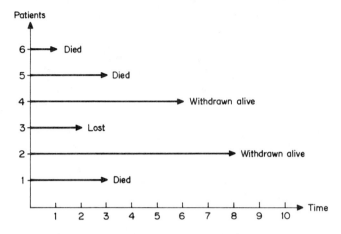

Figure 3.2. Clinical trial of six patients.

lifetimes are greater than $X_{(r)}$. Here $X_{(1)}, X_{(2)}\ldots$ denote the ordered observations, $X_{(1)}$ being the smallest, $X_{(2)}$ the second smallest, and so on.

(iii) *Random censoring*: In clinical studies, patients enter the study at random, are lost to follow-up, and the study may be terminated at a given time. In this case, the observations are censored at random times. Random censoring includes Type I censoring.

Example 3.2: Consider a clinical trial in which 6 patients are studied. Figure 3.1 provides their time of entry and final end point. The first patient enters at time 0 and dies at time 3. The second patient enters at time 1 and is alive at the end of the study, time 10. The third patient enters the study at time 2 and is lost to follow-up at time 4. The fourth patient enters at time 3 and is alive at time 10. Similarly, patients 5 and 6 are shown on the graph. The graph is now drawn in Figure 3.2 to have each patient start at 0.

Estimation of λ under censoring

(a) Under Type I censoring, let n_u be the number of uncensored observations and let the censoring time be T. The total amount of lifetimes observed is:

$$V = \sum_{i=1}^{n_u} X_i + (n - n_u) T$$

Since the parameter λ of the exponential distributions is estimated by the inverse of the average lifetimes in the censored case, adjustment to the average is to be made for incomplete observations. The equivalent of average in Type I censoring is the average of total lifetimes of censored as well as uncensored observations, based on the uncensored observations n_u. So the estimate of λ is given by:

$$\hat{\lambda}_I = \frac{n_u}{V} \qquad (3.6)$$

(b) The total amount of lifetimes observed in the case of Type II censoring is obtained by assuming that the remaining $n - r$ lifetimes are at least equal to the $X_{(r)}$, the largest among r observations. Then we have:

$$W = \sum_{i=1}^{r} X_{(i)} + (n - r)X_{(r)}.$$

The estimate of λ is again obtained with the help of the average of W based on r observations. Hence the estimate of λ in Type II censoring is given by:

$$\hat{\lambda}_{II} = \frac{r}{W} \qquad (3.7)$$

The probability distribution of $2\hat{\lambda}_{II} W$ is chi-square with $2r$ degrees of freedom, and we can use this fact to provide confidence intervals for λ.

The $(1 - \alpha)$-level confidence interval for λ under Type II censoring is:

$$\frac{X^2_{2r,\alpha/2}}{2W} < \lambda < \frac{X^2_{2r,1-\alpha/2}}{2W} \qquad (3.8)$$

(c) Let n_u be the number of observations that are not censored and let $U = \sum_{i=1}^{n} X_i$ be the total of censored as well as uncensored observations. Then, as in the case of Type I and II censoring, we have for Type III censoring:

$$\hat{\lambda}_R = \frac{n_u}{U} \qquad (3.9)$$

Example 3.3: The convention for a censored observation is to add a plus sign after them. For data given in Example 3.2, patients 2 and 4 are censored and patient 3 is lost to follow-up. The observations are $3, 9+, 2+, 7+, 3$, and 1. In this case:

$$\hat{\lambda}_R = \frac{3}{25} = .12$$

Many other parametric models are used in survival analysis, among them models based on gamma, Weibul, Raleigh, and lognormal distributions. The expressions of density functions and forces of mortality for these cases are complicated and are beyond the scope of this book.

Exercises

1. The times of death of 10 patients suffering from liver cancer over a one-year

period at a hospital were found to be (in months): 11.5, 1.5, 2, 3, 4, 2, 7.5, 8, 5, 9. Assuming a constant force of mortality, estimate its value.

2. Service times on the life of a battery are given below with the total number of failures during a given period.

Interval (month)	Number of batteries received for service	Number of failure
1	900	2
2	716	1
3	512	0
4	300	3
5	180	5
6	90	1
7	60	0
8	6	0

Assuming that the times are exponentially distributed, find the lifetime of the battery.

3. The breakdown time of an insulating fluid at 30 KV voltage is assumed to be exponential. Times (in seconds) are given on 8 units: 1, 2, 12, 25, 56, 68, 110, 125. Find the estimate of λ. What is the probability that the unit will not break before 150 seconds?

Follow-up Studies

Cause-and-effect relationships generally use *follow-up* studies. A forward-looking follow-up study is called *prospective*. Suppose we want to study the rates of death from lung cancer among smokers and nonsmokers. A *prospective follow-up* study will consist of a cohort of smokers and nonsmokers who are being followed up for a period of time. The number of deaths between the two groups will be an important factor in determining that smoking is a *cause* of lung cancer.

A *retrospective follow-up study* considers a cohort of patients with lung cancers and others without lung cancer, but it is followed up backward in time to ascertain the number of smokers and nonsmokers among them. Life-table techniques are commonly used in the analyses of data arising from such studies.

Suppose the risk factor such for smoking is denoted by C and the absence of the risk factor by $\bar{C}$. Also, let A denote the presence of the disease, such as lung cancer, and A its absence. Suppose the counts are given in the following 2×2 table.

Disease

	A	$\bar{A}$	
C	n_{11}	n_{12}	$n_{1.}$
$\bar{C}$	n_{21}	n_{22}	$n_{2.}$
	$n_{.1}$	$n_{.2}$	n

Risk factor

Study Designs

Consider the following study designs.

(i) *n fixed.* When the sample is such that n_{ij} are obtained without any restriction except that $n = n_{11} + n_{12} + n_{21} + n_{22}$, we have a *cross-sectional design* of the study. It is clearly seen that the random variables n_{ij} have a multinomial distribution, since the total number n is fixed and the number of observations in the four cells is random.

(ii) Suppose $n_{.1}$ and $n_{.2}$ are fixed. That is, the number of total cases having disease A and not having disease A is fixed. Because we include the presence and absence of risk factors, we have a *retrospective study design*. Notice that in such a case each of n_{11} and n_{12} are binomially distributed.

(iii) When we assume that $n_{1.}$ and $n_{2.}$ are fixed—that is, we fix those with risk factors C and $\bar{C}$ and observe n_{11} and n_{21} for the presence of disease A, we have a *prospective study design*. Here n_{11}, n_{21} are binomially distributed.

Another important quantity used in epidemiological studies is *relative risk*. Suppose the probability that an individual has disease A given that the individual is exposed to risk factor C is P_1 and, similarly, the probability that the individual has disease A given that he is not exposed to risk factor C is P_2. The *relative risk* is defined by:

$$R = \frac{P_1}{P_2}$$

The estimate of R from the above table can be given by:

$$\hat{R} = \frac{n_{11}/n_{1.}}{n_{21}/n_{2.}} \tag{3.10}$$

This is the relative risk of getting disease A when risk factor C is present versus when risk factor C is absent.

Another commonly used quantity is the *odds ratio*. This is the ratio of odds P_1/Q_1 and P_2/Q_2, where $Q_1 = 1 - P_1$ and $Q_2 = 1 - P_2$.

Odds are commonly used to express the probability of an event. Rather than saying that the probability of an event is .3, we say that the odds for the occurrence of the event are 3 to 7. The ratio of the odds reflects the relative importance of the occurrence of a disease in the presence and absence of a risk factor.

The odds ratio is:

$$\theta = \frac{P_1 Q_2}{P_2 Q_1}$$

The estimate of θ is:

$$\frac{n_{11}/n_{1.}}{n_{21}/n_{2.}} \times \frac{n_{22}/n_{2.}}{n_{21}/n_{1.}}$$

giving

$$\hat{\theta} = \frac{n_{11} n_{22}}{n_{12} n_{21}} \tag{3.11}$$

Example 3.4: Data were obtained to study a possible association between tonsillectomy (T) and infectious mononucleosis (IM). See Miller (1980).

	T	$\bar{T}$	
IM	44	159	203
Control	409	580	989
	453	739	1192

The estimate of the odds ratio of IM versus no IM is given by:

$$\hat{\theta} = \frac{44 \times 580}{409 \times 159} = .392$$

The confidence interval for the odds ratio can be found in Fleiss (1973). Comparison of several odds ratios also has received much attention in recent literature. For a recent survey and new results, see Zacks ad Soloman (1976).

In follow up studies, individuals get "lost" for various reasons. An individual may move to an unknown location for employment or other reasons. It is known that these individuals are alive at the time of being lost. However, there are also some individuals we know nothing about, and we regard them as "withdrawn" from the study.

An index called T-*year survival rate*, where the follow up takes place for T years, is used in actuarial studies and to compare treatments in clinical trials. Consider a cohort in a one-year interval $(x, x + 1)$. Let:

u_x = number of persons lost to follow up in the interval
w_x = number of persons withdrawn alive over $(x, x + 1)$
l_x = number of survivors at the beginning of the interval
d_x = number of deaths in the interval

Let l'_x be the number of persons exposed to the risk of death in the interval. Then:

$$l'_x = l_x = \tfrac{1}{2}(u_x + w_x)$$

Let q_z be the probability of death in the interval, $q_x = 1 - p_x$.

$$\hat{q}_x = d_x/l'_x$$

Then the T-year survival rate is denoted by $S(T)$ where its estimate is given by $S(T) = (1 - q_1)\ldots(1 - q_T)$.

$$\hat{S}(T) = (1 - \hat{q}_1)(1 - \hat{q}_2)\ldots(1 - \hat{q}_T)$$

or

$$\hat{S}(T) = \hat{p}_1\hat{p}_2\ldots\hat{p}_T \tag{3.12}$$

The variance of the T-year survival rate is estimated approximately by:

$$\hat{\text{Var}}(\hat{S}(T)) \approx \hat{S}^2(T) \sum_{x=1}^{T} (\hat{q}_x/l'_x\hat{p}_x) \tag{3.13}$$

Example 3.5: Deaths, survival, and withdrawal are studied for 126 male Connecticut residents with kidney cancer during the period 1946–51. The follow-up began in 1946. The data are given by Cutler and Ederer (1958).

$(x, x + 1)$	l_x	d_x	u_x	w_x
0–1	126	47	4	15
1–2	60	5	6	11
2–3	38	2	0	15
3–4	21	2	2	7
4–5	10	0	0	6

To calculate a 5-year survival rate for kidney cancer, we need l'_x q_x, p_x... as given below:

$(x, x+1)$	l'_x	$\hat{q}_x$	$\hat{p}_x$	$\hat{S}_{(x)}$
0–1	116.5	.40	.60	.60
1–2	51.5	.10	.90	.54
2–3	30.5	.07	.93	.50
3–4	16.5	.12	.88	.44
4–5	7.0	.00	1.00	.44

So the five-year survival rate is .44 from Equation (3.12).

The variance of the survival rate is obtained from Equation (3.13) as follows:

$$\hat{\mathrm{V}}\mathrm{ar}\,\hat{S}(4) = (.44)^2 \left[\frac{.4}{116.5 \times .6} + \frac{.1}{51.5 \times .9} + \frac{.07}{30.5 \times .93} + \frac{.12}{16.5 \times .88} + \frac{0}{7 \times 1} \right]$$

$$= (.1936)[.005722 + .002158 + .002468 + .008264]$$

$$= .0036$$

Exercises

4. In a cohort of 76 persons followed over a period of four years, the following data were obtained:

$(x, x+1)$	l_x	d_x	u_x	w_x
0–1	76	21	2	3
1–2	50	10	1	2
2–3	37	14	2	5
3–4	16	10	0	2

(a) Estimate the prevalence rate of the disease in every year
(b) Estimate the incidence rate for the four-year period
(c) Estimate the 4-year survival rate
(d) Estimate the variance of the survival rate

5. A study to discover the association between AIDs and homosexuality will be conducted at a hospital using patient records. Describe how to conduct:

(i) a retrospective study
(ii) a prospective study
(iii) a cross-sectional study

6. A study is being done to exhibit an association between a low incidence of the common cold and the use of vitamin C. A retrospective study design was used to give the data:

	Used vitamin C	Did not use Vitamin C
Had cold	5	20
Did not have cold	25	10

What do you conclude? Next, find the odds-ratio estimate and the estimate of the relative risk of catching a cold.

Product Limit (Kapln-Meier) Estimate

The method for estimating survival rates given by Kaplan and Meier (1958) is similar to the actuarial method. However, in this case the length of intervals is not assumed to be the same. The observations are ordered.

Let $X_1 \leq X_2 \leq \cdots \leq X_i$ be the times of death of i patients in an increasing order. We assume that if a death or loss occurs at the same time, the death will be recorded slightly before and the loss slightly after the time of occurrence. Suppose further that at time X_i, n_i patients are alive, including the one who died at X_i.

At time x_1, n_1 patients are alive and one death has occurred at time x_1. So the probability of survival at x_1 is:

$$\frac{n_1 - 1}{n_1}$$

The probability of survival at time x_2 is conditioned on the fact that there were n_1 survivors at time x_2. The conditional probability of survival at x_2 (given that there were n_1 survivors at x_1), since there are n_2 survivors at x_2, is obviously:

$$\frac{n_2 - 1}{n_2}$$

The unconditional probability of survival at x_2 is obtained as the product of the above two, using the formula for the two events A and B:

$$P(A \cap B) = P(A|B)P(B)$$

Hence the survival probability at x_2 is given by:

$$\hat{S}(x_2) = \frac{n_1 - 1}{n_1} \cdot \frac{n_2 - 1}{n_2}$$

In general, the *product limit estimate* or *Kaplan-Meier estimate* of survival at x_i is given by:

$$\hat{S}(x_i) = \left(\frac{n_1 - 1}{n_1}\right)\left(\frac{n_2 - 1}{n_2}\right) \cdots \left(\frac{n_i - 1}{n_i}\right) \tag{3.14}$$

If no losses occur in this case, the estimate is an ordinary binomial estimate.

Example 3.6: Consider a study in which 8 patients are observed with times of death given by 1, 2, 3, 5, and 7 months for 5 patients, and 3 patients are known to have been lost to follow up. One was lost between 2 and 4, another between 4 and 6, and the third between 6 and 8. The data are:

$$1, \quad 2, \quad 2+, \quad 3, \quad 3+, \quad 5, \quad 5+, \quad 7$$

Also:

$$n_1 = 8, \quad n_2 = 7, \quad n_3 = 5, \quad n_4 = 3, \quad n_5 = 1$$

with

$$X_{(1)} = 1, \quad X_{(2)} = 2, \quad X_{(3)} = 3, \quad X_{(4)} = 5 \quad \text{and} \quad X_{(5)} = 7$$

We then have the following estimates of survival function:

$$\hat{S}_{(0)} = 1$$

$$\hat{S}_{(1)} = \left(\frac{8-1}{8}\right) = .875$$

$$\hat{S}_{(2)} = \left(\frac{8-1}{8}\right)\left(\frac{7-1}{7}\right) = .75$$

$$\hat{S}_{(3)} = \left(\frac{8-1}{8}\right)\left(\frac{7-1}{7}\right)\left(\frac{5-1}{5}\right) = .6$$

$$\hat{S}_{(5)} = \left(\frac{8-1}{8}\right)\left(\frac{7-1}{7}\right)\left(\frac{5-1}{5}\right)\left(\frac{3-1}{3}\right) = .4$$

$$\hat{S}_{(7)} = 0$$

The survival function is graphed in Figure 3.3.

Variance of $\hat{S}(x)$. The approximate estimate of the variance of $\hat{S}(x)$ at $x_{(i)}$ is given by:

$$\hat{V}ar(\hat{S}(x_{(i)})) = [\hat{S}(x_{(i)})]^2 \sum_{j=1}^{i} \frac{1}{n_j(n_j - 1)} \tag{3.15}$$

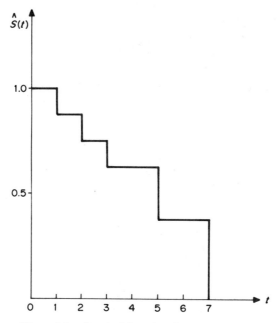

Figure 3.3. Survival function for Example 3.6.

For Example 3.6, the estimate of variances are:

$$\hat{V}ar(\hat{S}(1)) = \left(\frac{8-1}{8}\right)^2 \frac{1}{8(8-1)} = .01$$

$$\hat{V}ar(\hat{S}(2)) = (.75)^2 \left[\frac{1}{8(8-1)} + \frac{1}{7(7-1)}\right]$$

$$= .0234$$

Kaplan-Meier estimates are used in many important studies. To apply it to Stanford Heart Transplant data, see Turnbull and Brown (1974).

Exercises

7. For a clinical study of 6 patients, the times of death are 1, 3, 3 +, 5, 6, and 7. Find and graph the product limit estimate of the survival function.

8. The length of times of remission in acute myelogeneous leukemia under the maintenance chemotherapy for 11 patients are given as 9, 13, 13 +, 18, 23, 28 +, 31, 34, 45 +, 48, and 161 +. Find the product limit estimate $\hat{S}_{(45)}$ and its variance. The data are given by Miller (1981).

9. The survival times of 81 patients with melanoma are given below. Arrange the data in increasing order and find the estimate of the survival function at $t = 16$. What is its variance?

136	193 +	34	50 +
194 +	134 +	38	21 +
54	125 +	46	76 +
103	32	60	65
81 +	140 +	31	124
132 +	93 +	23	50
118	19	213 +	34
26	181 +	110	148 +
80	20	65	100 +
58	138	158	114 +
14	147 +	85	44 +
98	151 +	37	13 +
59	130 +	86 +	234
152 +	66	25	108 +
140 +	80 +	190 +	26
53	38	215 +	57
102	21	67 +	27
73 +	130	40	19
55 +	141	27	124 +
90	162 +	129 +	23
			16 +

Source: Susarla, V., and Van Ryzin, J., (1978).

Mantel-Haenszel Test

We discussed 2×2 contingency tables in Chapter 7, Volume I, where tests of independence of attributes leading to contingency tables are developed. In survival analysis, contingency tables arise naturally. Suppose two treatment groups of patients are being observed for survival during a given year. The number of deaths and survivals in each group will provide a contingency table. The test of the hypothesis that the probabilities of survival under the two treatments are equal, is equivalent to the test of independence in the contingency tables and uses a chi-square statistic with one degree of freedom. The Mantel-Haenszel test is concerned with testing the equality of these probabilities in several tables simultaneously.

Consider the following table:

Treatments	I	II	
Number of deaths	a	b	m_1
Number of survivals	c	d	m_2
	n_1	n_2	n

.istic is:

$$\chi^2 = \frac{n(ad - bc)^2}{n_1 m_1 n_2 m_2}$$

.ier n_1, n_2, m_1, m_2 as fixed. Then the table has only one random variable
$\;$ .ose value is a. The random variable A has a hypergeometric distribution.
We see from Chapter 4, Volume I that the mean and variance of the
hypergeometric random variable are given by:

$$E(A) = \frac{n_1 m_1}{n}$$

and

$$\text{Var}(A) = \frac{n_1 m_1 n_2 m_2}{n^2(n - 1)}$$

Then the χ^2 statistic can be written as:

$$\chi^2 = \frac{n}{n - 1} \frac{\left(a - \dfrac{n_1 m_1}{n}\right)^2}{\dfrac{n_1 m_1 n_2 m_2}{n^2(n - 1)}} = \frac{n}{n - 1}\left[\frac{a - E(A)}{\sqrt{\text{Var}(A)}}\right]^2$$

The coefficient $\dfrac{n}{n - 1}$ is the correction of bias in $\text{Var}(A)$. Hence the test can be
based equivalently on $[a - E(A)]/\sqrt{\text{Var}(A)}$, which has a standard normal
distribution.

Suppose now that several 2×2 tables are given and it is not appropriate to
pool the tables. For example, the data on treatments I and II have come from
several, say k, hospitals that are not homogeneous. The k tables are given
below:

a_1	b_1	m_{11}
c_1	d_1	m_{12}
n_{11}	n_{12}	n_1

a_2	b_2	m_{21}
c_2	d_2	m_{22}
n_{21}	n_{22}	n_2
$\vdots$	$\vdots$	$\vdots$
a_k	b_k	m_{k1}
c_k	d_k	m_{k2}
n_{k1}	n_{k2}	n_k

The Mantel-Haenszel statistic is:

$$MH = \frac{\sum_{i=1}^{k}(a_i - E(A_i))}{\sum_{i=1}^{k}\sqrt{\text{Var } A_i}}$$

If a correction is needed, we have:

$$MH_c = \frac{\left|\sum_{i=1}^{k}(a_i - E(A_i))\right| - \frac{1}{2}}{\sum_{i=1}^{k}\sqrt{\text{Var}(A_i)}}$$

The statistic has a standard normal distribution (with or without correction) and gives the test of equality of proportions.

Example 3.7: To discover a possible relationship between tonsillectomy (I) and infectious mononucleosis (IM), data from Stanford University's Cowell Student Health Center are given by Miller et al. (1980). The data are by ages and cells contain frequencies.

Ages	18		19		20		21	
	T	$\bar{T}$	T	$\bar{T}$	T	$\bar{T}$	T	$\bar{T}$
IM	6	17	3	39	12	29	8	38
$\overline{IM}$	17	32	26	70	34	78	48	91

	22		23		24	
	T	$\bar{T}$	T	$\bar{T}$	T	$\bar{T}$
	5	10	2	7	4	5
	45	73	29	37	36	39

To test the hypothesis that the proportion of tonsillectomies in the infectious mononucleosis group and the control group ($\overline{IM}$) are equal for all ages, we use the Mantel-Haenszel statistic. The data are arranged as in Table 3.1.

Table 3.1 Data

Age	a	n_1	n_2	m_1	m_2	n
18	6	23	49	23	49	72
19	3	42	96	29	109	138
20	12	41	112	46	107	153
21	8	46	139	56	129	185
22	5	15	118	50	83	133
23	2	9	66	31	44	75
24	4	9	75	40	44	84

Table 3.2 Calculations for the MH Statistic

a	$E(A) = \dfrac{n_1 m_1}{n}$	$a - E(A)$	$n^2(n-1)$	$\mathrm{Var}(A) = \dfrac{n_1 m_1 n_2 m_2}{n^2(n-1)}$
6	7.347	−1.347	368,064	3.451
3	8.826	−5.826	2,609,028	4.885
12	12.327	−.327	3,558,168	6.352
8	13.924	−5.924	6,297,400	7.335
5	5.639	−.639	2,334,948	3.146
2	3.720	−1.720	416,250	1.946
4	4.286	−.286	585,648	2.029
Total		−16.069		29.144

The equations are:

$$MH = \frac{-16.069}{\sqrt{29.144}} = -2.977$$

$$MH_c = \frac{|-16.069| - .5}{\sqrt{29.144}} = 2.88$$

Since the standard normal table gives $z_{.99} = 2.33$, the hypothesis of the equality of proportions is rejected at .01 level. That is, there is relationship between tonsillectomy and infectious mononucleosis. The relative risks also can be obtained in this case.

Other nonparametric procedures as discussed in Chapter 8, Volume I, can be applied to survival data. For example, if one wants to test the hypothesis that two survival distributions are the same, the Kolmogorov-Smirnov test can be used. The percentage points of the two-sample Kolmogorov-Smirnov statistic are given in Table XIV in the Appendix.

Exercises

10. A drug (D) is tested to see its effectiveness in reducing vomiting (V) after surgery in three hospitals. Given the following data, test the relationship at .01 level using the MH statistic ($\bar{D}$ = placebo):

	Hospital					
	I		II		III	
	V	$\bar{V}$	V	$\bar{V}$	V	V
D	120	30	30	10	60	20
$\bar{D}$	10	90	15	5	30	16

11. Two surgical procedures (I and II) are used in open-heart surgery by four different surgeons. Their experience is given in terms of their successes (S) and failures (F) over a period of several years. Test the hypothesis of equality of the success rates for the surgeons using the MH statistic:

	Surgeon							
	1		2		3		4	
	I	II	I	II	I	II	I	II
S	5	2	7	1	3	0	10	2
F	10	2	0	2	1	2	2	3

12. To study the relationship between compaigning by the incumbent President (P) in the election district of a congressman and the number of votes cast, the following data were obtained for urban and rural areas after elections:

	Rural				Urban	
	P	$\bar{P}$			P	$\bar{P}$
Favored	1370	1450		Favored	1250	107
Not favored	170	150		Not favored	117	37

Test for an association at the .05 level using the MH statistic.

Chapter Exercises

1. The data on mortality of 43 patients with granulocytic leukemia are given in Chapter Exercise 3 in Chapter 2. Assume that the times of death are exponentially distributed. Find the estimate of the survival function. Estimate the probability that:

   (i) A patient with granulocytic leukemia will survive more than 730 days
   (ii) The patient will die before 200 days
   (iii) The patient will live between 500 to 1,000 days

2. Suppose the data in Example 15.7 are accumulated in one table giving the number of students with tonsillectomy and infectious mononucleosis. Find the relative risk of infectious mononeucleosis. Calculate the odds ratio.

3. The survival times of male mice exposed to radiation are given for two groups, one fed lactobacillus and one not fed lactobacillus. Use the Kolmogorov-Smirnov two-sample test to find if their survival experiences differ at .10 level of significance.

Group *I* (weeks)					Group *II* (weeks)				
152,	109,	135,	57,	71,	132,	140,	91,	85,	141
87,	83,	125,	128,	136	158,	119,	145,	47,	82

Summary

The *prevalence rate* and the *incidence rate* for a disease are commonly used in epidemiology. An important concept in survival analysis is the *survival function*, which gives the probability of survival at age x. For the exponential model, it is given by:

$$S(x) = e^{-\lambda x}$$

The data in survival studies are usually *censored. Type I* and *Type II* censoring mechanisms are applied in engineering; *random censoring* is usually done in survival studies. *Follow up studies* can be *retrospective* or *prospective*. Such studies lead to the calculation of the *relative risk* and *odds ratio*. Another actuarial measure used in survival studies is the T-year survival rate, which can be calculated from life tables. Commonly used techniques in recent years are those of the *product limit (Kaplan-Meier)* estimate of survival probability and the *Mantel-Haenszel test* for testing the hypothesis or lack of association when data are obtained in several 2×2 contingency tables that are not homogeneous. Usual nonparametric procedures can be applied to survival data when assumptions of parametric models are not appropriate.

References

Benedetti, Jacqueline K., et al. Effective sample size for tests of censored survival data, *Biometrika* 1982, *62*. 343–49.

Cox, D. R. Regression models and life tables, *Journal of the Royal Statistical Society*, 1972, *B34*. 187–220.

Cutler, S. J., and Ederer, F. Maximum utilization of life table method in analyzing survival, *J. of Chronic Diseases*, 1958, *8*, 699–712.

Elandt-Johnson, Regina C., and Johnson, Norman, L. *Survival Models and Data Analysis*. New York: John Wiley & Sons, 1980.

Fleiss, J. L. *Statistical Methods for Rates and Proportions*. New York: John Wiley & Sons, 1973.

Gross, Alan J., and Clark, Virginia A. *Survival distributions: Reliability applications in the biomedical sciences*, New York: J. Wiley & Sons, 1975.

Kalbfleish, John D., and Prentice, Ross L. *The Statistical Analysis of Failure Time Data*. New York: John Wiley & Sons, 1980.

Kaplan, E. L., and Meier, P. Nonparametric estimation from incomplete observations, *J. Am. Statist. Assoc.*, 1958, *53*, 457–81.

Mantel, N., and Haenszel, W. Statistical aspects of the analysis of data from restrospective study of disease, *J. National Cancer Institute*, 1959, *22*, 719.

Miller, Rupert G., Jr.; Efron, Bradley; Brown, Byron W., Jr.; and Moses, Lincoln E. *Biostatistics Casebook*. New York: John Wiley & Sons, 1980.

———. *Survival Analysis*. New York: John Wiley & Sons, 1981.

———, and Halpern, Jerry. Regression with survival data, *Biometrika*, 1982, *69*, 521–31.

Susarla, V., and Van Ryzin, J., Empirical Bayes estimation of a distribution (survival) function from right censored observations, *Ann. Statint.*, 1978, *6*, 740–54.

Turnbull, B.; Brown, B. W.; and Hu, M. Survivorship analysis of heart transplant data, *J. Am. Statist. Assoc.*, 1974, *69*, 74–80.

Zacks, S., and Solomon, H. On testing and estimating the interaction between treatments and environmental conditions in binomial experiments: The case of two stations, *Commun. Statist.- Theo. Meth.*, 1976, *A5(3)* 197–223.

chapter four

Clinical Trials

In testing new drugs or new medical and surgical treatment methods on human subjects, field trials known as clinical trials are made. Clinical trials are conducted to compare at least two interventions, one of which is usually the standard or control intervention. The subjects are assigned to these interventions at random. A properly designed clinical trial is extremely important for drawing valid inferences from its results. The use of randomization in clinical trials is of recent origin. Objections to randomization are usually made by clinicians for practical as well as ethical reasons. Experiments on human subjects, such as in environmental health or cancer therapy, are necessary to understand the effect of many chemical substances and other therapeutic procedures. Controversies such as radical mastectomy versus chemotherapy as treatment for breast cancer cannot be easily solved without extensive clinical trials.

Professional medical societies and government agencies such as the Food and Drug Administration and the National Institutes of Health have strict regulations for conducting clinical trials, as do most countries of the world. Designing clinical trials has become an important area of study in experimental therapeutics and in studies of survival. The design of experiments has been discussed earlier in Chapter 11, Volume I. However, new problems of experimental designs appear in clinical trials, since the experimental units are human subjects. For example, in most situations, informed consent of the subject is required by law. Such interference in experimentation makes randomization difficult to implement. New approaches to experimentation in human subjects, such as double-blind studies, have been developed to solve such problems. Objections to blinding the treatment for the physician are made on ethical and practical grounds in view of complications, and in many clinical trials blinding is difficult. Clinical trials are expensive and in diseases like cancer, they are performed over a long period of time. One of the earliest large-scale field trials was for testing the efficacy of Salk polio vaccine in 1952.

In double-blind experiments, treatments are coded so that neither the patient nor the physician knows the identity of the treatment. Such plans have been criticized on practical basis, since in case of complications, the identity of the treatment is not found easily. Such plans are widely used in cases where drugs are unlikely to cause complications. However, it is not always possible to implement double-blind clinical trials.

Randomization is attacked on ethical and moral grounds. The experimentation requires that some subjects many not get the best possible treatment. Also, if the new treatment proves to be worse than the standard treatment, the physician would not be happy with the results. To solve this difficulty, sequential medical trials can be used so that the inferior treatment can be eliminated as soon as possible.

The three basic principles of ethics in the conduct of clinical trials have been proposed by Lebacqz (1983). First is respect for the patient. This can be accomplished in part by informed consent. Second, there should be some benefit for the patient. There should be a clear indication that the new treatment is a better treatment and is likely to benefit the patient. Third, there should be justice to the individual. This principle implies that there is equal distribution of burdens and benefits, and no one will be adversely affected on the whole by the clinical trial.

Many physicians view each episode of treating a patient as an experiment, since the information so obtained is used by the physician for treating future patients. In this way, the physician is performing a clinical trial. However, it is not a well-designed one, since as experimental plan is not made in such cases. This chapter describes common methods used in clinical trials. Statistical procedures usually applied to data obtained from clinical trials are those of survival analysis and life-table techniques, discussed in Chapters 2 and 3. The importance of controlled clinical trials in modern medicine has resulted in the regular publication of a journal, *Controlled Clinical Trials*, which provides latest results in the area. The literature on clinical trials is vast. See Shapiro and Louis, eds. (1983) and other publications listed at the end of the chapter.

Experimental Designs

Several kinds of disigns are used in randomized clinical trials.

(i) *Complete randomization*: The simplest clinical trial is concerned with the measurement of efficacy of a new drug versus a standard drug or the efficacy of a new drug versus a placebo. A *placebo* is an identical-looking drug that has no active ingredients. In such a trial, the study may be *self-controlled* or *externally controlled*. In a self-controlled study, a single treatment is tried with a control in the same patient. The control and treatment observations are taken over widely separated intervals so that there is no dependence among them. In an externally controlled study, the control is external to ths subject. Sometimes *historical controls* are used for this purpose. For groups of patients who are terminally ill, historical controls are logical. In this case, we know the course of the events. That is, unless treated, every patient will die. Historical controls with a randomized trial selecting patients at random for treatment are enough for a valid clinical evaluation of the new treatment in such cases.

The design of a *completely randomized* trial requires that the subjects be assigned at random to a placebo (*A*) or a drug (*B*). The randomization can be accomplished through a table of random numbers. Such a design may lead to unequal number of subjects for *A* and *B*.

Analysis of data from the above trail can be done by choosing parametric models. Suppose we assume that the responses are normally distributed. Then the one-sample or two-sample *t*-test can be used as the case requires. If several drugs are to be compared simultaneously, we use analysis-of-variance techniques. When the number of observations is small or we cannot assume a parametric model, techniques from nonparametric statistics are used. Chapter 8 provides most of these methods.

(ii) *Standard Experimental Designs*: In some clinical trials, several drugs such as pain relievers, tablets, or ointment can be tested on the same individual. Assuming that the observations at various times are independent of each other, we can use the *randomized block design*, where the individual acts as a block and treatments are assigned at random. *Latin square* designs can similarly be used. Statistical analysis of the data will be done through the analysis of variance as discussed in Chapter 10. When the various treatments on the individual are correlated, *crossover designs* are used.

(iii) *Crossover Designs*: Assume that we have treatments *A*, *B*, and *C* for experimentation. Then we have the following six different arrangements for the treatments.

(1)	A	B	C
(2)	A	C	B
(3)	B	C	A
(4)	B	A	C
(5)	C	A	B
(6)	C	B	A

A crossover design requires that each arrangement be given to an individual. The above design does not allow the repeated effect of the same treatment on a subject to be evaluated. A modification proposed is the following design, where six treatments are given to an individual. Again the arrangements given below are assigned at random. We note that all three treatments find at least one repetition on a given individual.

(1)	A	B	C	C	A	B
(2)	B	C	A	A	C	B
(3)	C	A	B	B	A	C

In the above design, using at least three subjects, all orders such as A followed by B and A followed by C, as well as A followed by A, and so on can be tested.

In a *double-crossover design*, all possible pairs are repeated in individuals. An example is given below, where two treatments A and B are repeated. Note that AA, BB, AB, and BA are repeated in addition to repeating A and B. We need six individuals for this crossover design.

(1)	A	A	B	B
(2)	B	B	A	A
(3)	A	B	B	A
(4)	B	A	A	B
(5)	A	B	A	B
(6)	B	A	B	A

(iv) *Play-the-winner rule*: When two treatments are being compared, the play-the-winner rule requires that patients be assigned to one treatment chosen at random and continue to receive the same treatment until it fails to cure. Then the patients are assigned to the second treatment until it fails to cure. The assignment of treatment continues back and forth in this way. This procedure is a special case of a sequential experiment, where the result of the experiment is needed in order to assign an observation. Some sequential medical trials are discussed later in the chapter.

Example 4.1: A randomized clinical trial (externally controlled) was conducted to study the effectiveness of tetanus antitoxin. The data reported by Armitage and Gehan (1974) are given in Table 4.1.

Thee estimates of the probability of effectiveness of the antitoxin are obtained below. Recall that $\text{Var}(\hat{p}) = p(1-p)/n$ and $\hat{\text{Var}}\,\hat{p} = \hat{p}(1-\hat{p})/n$, and the standard deviation of $\hat{p} = \sqrt{\hat{\text{Var}}\,\hat{p}}$.

$$\hat{p}_{\text{treated}} = \frac{21}{41} = .512, \text{ standard deviation} = .078$$

$$\hat{p}_{\text{control}} = \frac{9}{38} = .237, \text{ standard deviation} = .069$$

$$\hat{p}_{\text{treated}} - \hat{p}_{\text{control}} = .275, \text{ standard deviation} = .104$$

Table 4.1 Data

	Number of patients	Number of survivals
Treated	41	21
Control	38	9
Total	79	30

Exercises

1. A study was made to compare two soporific drugs used on the same individual. Data are given for the additional hours of sleep gained by the use of the drugs on 10 patients:

Patients

	1	2	3	4	5	6	7	8	9	10
A	.7	− 1.6	− .2	− 1.2	− .1	3.4	3.7	.5	0	2
B	1.9	.8	1.1	.1	− .1	4.4	5.5	1.6	4.6	3.4

Source: Fisher (1946) and quoted by Meier (1983).

Test the hypothesis at .05 level of significance that the drugs are equally effective.

2. A randomized clinical trial to measure the effect of a new drug on the survival (in months) from a terminal disease was conducted with the results:

New drug: 7,   3,   5 +,   8,   12,   9
Standard drug: 8,   17,   2,   3,   7 +,   6

What can you say about the new treatment?

3. Suppose the data in Table 4.1 are further subdivided into two groups, high risk and low risk (a process of post stratification), in the tetanus experiment:

High risk

	Survival	Deaths	Total
Treatment	10	14	24
Control	4	26	30

Low Risk

	Survival	Deaths	Total
Treatment	11	2	13
Control	5	3	8

Source: Armitage and Gehan (1974).

How would you analyze these data further so as to see if the treatment has an effect?

Example of a Large Clinical Trial

Canner (1981) presents a Coronary Drug Project study as an example of a large-scale clinical trial.

(i) *Trial*: A randomized, double-blind, controlled, and multicenter clinical trial was conducted to evaluate the efficacy of lipid-influencing drugs on the long-term therapy of coronary heart disease.

(ii) *Drugs used*: The following five drugs were used.

    (i) Mixed conjugated equine estrogens (2.5 and 5.0 mg/day)

    (ii) Clofibrate (1.8g/day)

    (iii) Dextrothyroxine (6.0 mg/day)

    (iv) Nicotinic acid (3.0 g/day)

    (v) Lactose placebo

(iii) *Patient population*: From March 1966 to October 1969, 53 cooperating clinics participated, with a total of 2,789 patients in the placebo group and 8,341 in one of the five drug groups. To enter the study, a patient had to be a male between 30 and 64 years of age and he must have had ECG-documented evidence of myocardial infarction occurring not less than three months previously.

(iv) *Follow up*: Patients in the study visited the clinic every four months for 5 to 8.5 years.

(v) *Data and safety monitoring committee*: The trial was monitored by a committee of sixteen professionals, including seven clinicians, four statisticians, two pharmacologists, an epidemologist, a lipid chemist, and a clinical chemist. The committee met regularly to discuss matters connected with the trial.

(vi) *Early termination*: Because of an increased incidence of cardiovascular events, three of the five drugs were terminated. Only Clofibrate and nicotinic acid were followed to the end of the trial. Other details about this clinical trial are described by Canner. When only part of the trial is terminated, the statistical problems are complicated. However, when the whole trial is terminated early, statistical procedures for censored data can be used.

Statistical Considerations in Early termination

A clinical trial may be terminated earlier than planned for various reasons. When the trial is terminated at a time T, we have Type I censoring (see Chapter 3). When the trial is terminated as a result of having obtained a given number of failures (or deaths), we have Type II censoring.

Comparisons of two survival distributions under censoring can be made by using censored versions of well-known tests. For example, the *Kolmogorov-Smirnov statistic* (see Chapter 8) Volume I, gives a test of the hypothesis of equality of two distribution functions. The censored version of this test is called the *Tsao-Conover test*, which is applicable in the above situation.

When random censoring of observations takes place in clinical trials, we can resort to the Kaplan-Meier statistic to estimate the survival function as discussed in Chapter 3.

Multicenter Clinical Trial Data

When a clinical trial is conducted by several clinics, the data should not be pooled, since the clinics may differ in their makeup. The Mantel-Haenszel procedure is used to test the equality of proportions across clinics if the data are available in the form of 2×2 contingency tables.

Another method is based on the *P-values* (the *observed significance level*) from each data set. Consider k clinics and let

$$S = \text{the statistic for testing the hypothesis}$$
$$S_i = \text{observed value of the statistic from clinic } i$$

Let $P(S > S_i|H) = \alpha_i^0$ be the observed significance level P-value of the statistic for clinic i. Then

$$U = -2 \sum_{i=1}^{k} \log \alpha_i^0$$

has the chi-square distribution with $2k$ degrees of freedom. For large values of U, we reject the hypothesis overall at the given significance level α. For further discussion, see Elandt-Johnson and Johnson (1980).

Consider the case of three centers where clinical trials are held to compare two interventions. We assume that the interventions were compared using the t statistic. The P-values obtained are:

$$\alpha_1^0 = .03$$
$$\alpha_2^0 = .001$$
$$\alpha_3^0 = .11$$

Then:

$$U = -2[\log .03 + \log .001 + \log .11]$$
$$= -2[-1.52 - 3.00 - .96] = 10.96$$

The tabulated value $X_{6;90}^2 = 10.6$. Hence we reject the hypothesis of the equality of two interventions at a 10% level of significance.

Noncompliance of the Original Plan

Practical problems arise in clinical trials when original plans cannot be carried out. Suppose in a trial to compare a medical and surgical procedure it is

Table 4.2 European Coronary Surgery Data

	Assigned to Surgical Treatment		Assigned to Medical Treatment	
	Actual medical treatment	Actual surgical treatment	Actual medical treatment	Actual surgical treatment
Survivals	20	354	296	48
Deaths	6	15	27	2
Total patients	26	369	323	50
Death rates per hundred	23.1	4.1	8.4	4.0

Source: Armitage (1983).

discovered that after the patient is assigned to a medical procedure a surgical procedure is needed. Similarly, a patient assigned to a surgical procedure may need only medical treatment. Care is needed in analyzing the results of a trial.

Example 4.2: In coronary-artery-bypass surgery in stable angina pectoris conducted by the European Coronary Surgery Group in 1979, a clinical trial compared medical and surgical procedures. The data are given in Table 4.2. If we concentrate only on the group of adherents of a given procedure, the mortality rates for surgery (4.1%) seem much lower than for medical (8.4%). However, it is necessary to compare the death rates from all the assigned patients and *not* from the actually assigned patients. That is, we should compare the total assigned patients to the surgical procedure $(26 + 369 = 395)$ with the total assigned to the medical procedure $(50 + 323 = 373)$, since the policy of randomization was used for the whole group.

The death rates (per hundred) are then:

$$\text{Surgical} = \frac{(21)(100)}{395} = 395$$

$$\text{Medical} = \frac{(29)(100)}{373} = 7.8$$

The difference between the surgical (5.3%) and medical rates (7.8%) is not significant at level .05.

Sample-Size Determination

Problems of determining sample sizes in some simple cases were discussed in Chapters 6, Volume I. Sample-size determination is important in the case of clinical trials.

(i) *Estimation of a proportion from one sample.* Suppose a clinical trial uses historical controls. Therefore, the problem of estimating a death rate or a cure rate is reduced to an estimation of a proportion from the sample of the treatment. Assume, as before, that we require the estimate to be within d percent of the unknown proportion p, with a preassigned probability of $1 - \alpha$. From Equation (6.10) in Chapter 6, Volume I, we have

$$n = z_{\alpha/2}^2/4d^2 \tag{4.1}$$

where $z_{\alpha/2}$ is the $\alpha/2$-th percentile of the standard normal distribution. The closest integer to the right-hand side of equation (4.1) is the sample size.

(ii) *Estimation of the mean with given variance.* If the clinical trial provides a continuous response, such as an increase in blood pressure or a reduction in cholestrol level, then we assume it to be normally distributed with unknown mean μ and known variance σ^2.

Suppose a $(1 - \alpha)$-level confidence interval of the mean μ is required, whose length is $2d$. Then we know from Equation (6.9) Volume I that:

$$n = z_{\alpha/2}^2 \sigma^2/d^2 \tag{4.2}$$

(iii) *Comparison of two groups.* When we compare a control group and a treatment group, we are interested in the change between the means $\mu_t - \mu_c$, where μ_t is the mean of the treatment group and μ_c is the mean of the control group. We are usually interested in testing the hypothesis that the difference is μ_0 versus the difference is μ_1. Ordinarily, $\mu_0 = 0$. Let the test be done at level α and have power $1 - \beta$. Suppose the size of both groups is the same, denoted by n. Let σ_0^2 and σ_1^2 be the variances of the difference of the random variables measuring the treatment and control effects and z_α, z_β be the αth and βth percentiles of the normal distribution. Suppose that the rejection of the null hypothesis is based on the difference of the sample averages, say $\bar{z}$. Then under the null hypothesis, $\bar{z} \sim N(\mu_0, \sigma_0/\sqrt{n})$, and under the alternative $\bar{z} \sim N(\mu_1, \sigma_1/\sqrt{n})$. Hence the null hypothesis is rejected when

$$\bar{z} > \mu_0 - z_\alpha \frac{\sigma_0}{\sqrt{n}}, \quad \text{if} \quad \mu_1 < \mu_0$$

Since power of the test is $1 - \beta$, we have:

$$P\left\{\bar{z} > \mu_0 - z_\alpha \frac{\sigma_0}{\sqrt{n}} \middle| \mu_1, \sigma_1 \right\} = 1 - \beta$$

Hence

$$\mu_0 + z_\alpha \sigma_0/\sqrt{n} = \mu_1 + z_\beta \sigma_1/\sqrt{n}$$

giving

$$n(z_\alpha \sigma_0 + z_\beta \sigma_1)^2/(\mu_1 - \mu_0)^2 \tag{4.3}$$

For example, suppose $\alpha = .05 = \beta$. Let $\mu_0 = 0$ and $\mu_1 = 5$ with $\sigma_0 = 10$ and

$\sigma_1 = 15$. Then:

$$n = [(10)(1.645) + (15)(1.645)]^2/(5 - 0)^2$$
$$= 68$$

The clinical trial then has 136 subjects, with 68 for the control group and 68 for the treatment group.

(iv) *Paired observations*. If the clinical trial is performed such that the treatment and control observations are obtained on the same subject, we test the hypothesis that $\mu_0 = 0$ versus that the difference is μ_1. Let σ_d^2 be the variance of the difference. Using equation (4.3), with $\sigma_0 = \sigma_1 = \sigma_d$ and $\mu_0 = 0$, we have:

$$n = \frac{(z_\alpha + z_\beta)^2 \sigma_d^2}{\mu_1^2} \tag{4.4}$$

(v) *Comparing two proportions*. Here we consider the analogous problem of comparing proportions of two groups. Let the difference between the control and treatment proportions be p_0 versus that the difference is p_1. Let the significance level be α and have power $1 - \beta$. Assuming a large sample, the approximate distribution of the difference of proportions is normal with means p_0 and variance $p_0(1 - p_0)/n$ under the null hypothesis. Under the alternative hypothesis, the distribution is normal with mean p_1 and variance $p_1(1 - p_1)/n$. Using the same arguments as in deriving the sample size in case (a), we have:

$$n = \frac{(z_\alpha\sqrt{p_0(1 - p_0)} + z_\beta\sqrt{p_1(1 - p_1)})^2}{(p_1 - p_0)^2} \tag{4.5}$$

For example, let $p_0 = .01$ and $p_1 = .05$ with $\alpha = .05 = \beta$. Then the number of subjects per group is:

$$n = [(1.645)(.099) + (1.645)(.218)]^2/(.05 - .01)^2$$
$$= 170$$

Exercises

4. For a clinical trial using historical controls, an estimate of the proportion of cured by a treatment is required within .02 of the unknown proportion. The confidence level is required to be .95. Find the size of the clinical trial.

5. A trial to compare two treatments for mosquito bites is made such that both treatments can be applied to the same individual, say, the right and left hands, which are randomly assigned to the treatments. It is required to have significance level .05 and power .90 for detecting a difference between the

proportions cured of .1. The null hypothesis is that there is no difference between the treatments. Find the sample size.

Sequential Clinical Trials

In sequential clinical trials, the sample size is not fixed in advance. The result of each outcome is needed to decide whether we continue the experimentation or make a final decision. In many clinical trials, such results are not immediately available, since the effect of treatment may take a long time to occur. Sequential medical trials are special cases of sequential experimentation see Chapter 9.

Sequential experiments have two aspects. The first is the *stopping rule*, which provides a rule for stopping the experiment or continuing the experiment. The second is the *decision rule*, once the experiment has been stopped. When the experimenter decides not to take more than a given number of observations, the sequential plan is called *truncated* or *closed*. When there is no restriction on the sample size, the sequential plan is *nontruncated* or *open*.

Sequential medical trials have great potential in medical experimentation. For a fuller account beyond the few simple examples of sequential plans provided here see Armitage (1960).

Comparison of Two Treatments

(i) *Discrete case.* Consider a clinical trial in which we are testing the hypothesis that there is no preference for treatment A and B versus the hypothesis that one is better than the other. Let $\theta = P$ (A is preferred to B):

$$H: \quad \theta = .5$$
$$A: \quad \theta \neq .5$$

To make the alternative hypothesis simple, we can choose two points, $\theta = \theta_0$ or $\theta = \theta_1 = 1 - \theta_0$, using two numbers on either side of .5.

Let $y =$ the number of excess preferences of A over B. Then we can express the stopping rule and decision procedure with the help of a graph providing boundaries of the continuation region and the acceptance and rejection regions. At each stage, when observations are taken, we calculate the value of y and plot it to determine the next step, depending on where the value lies. The procedure arises from the sequential probability ratio test in Chapter 9.

To obtain the straight-line boundaries of the region, we need constants a_1, a_2 and b, which are given in Table 4.3 for a few values of θ_0. The probabilities of Type I and Type II errors are $\alpha = .05$ and $\beta = .05$.

The lower boundary of the rejection region for the null hypothesis and acceptance of $\theta = \theta_1$ is given by the line:

$$y = a_2 + bn$$

The upper boundary for acceptance of $\theta = \theta_0$ is given by the line:

$$y = -a_2 - bn$$

The upper and lower boundaries for the acceptance of the null hypothesis are given by:

$$y = a_1 + bn$$

and

$$y = a_1 - bn$$

In Figure 4.1, we give these regions and the continuation regions.

Table 4.3 also gives the expected number of observations saved by using the sequential sampling plans as compared to a fixed sampling plan having the same probabilities of the two kinds of errors. For other values, see Armitage (1960).

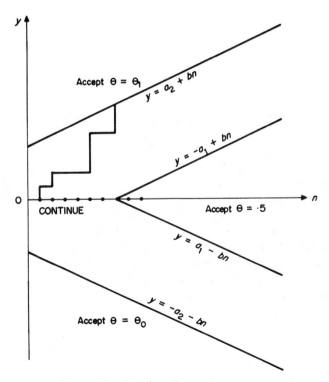

Figure 4.1. Continuation and acceptance regions.

Table 4.3 Open plan for $\alpha = .05 = \beta$

θ_0	θ_1	a_1	a_2	b	l	m	r
0.6	0.4	17.94	14.65	0.1007	215	160	319
0.8	0.2	5.25	4.29	0.3219	21	17	30
0.9	0.1	3.31	2.70	0.4650	10	9	14

Source: Armitage (1960) page 27.

l = approximate number of preferences required when the null hypothesis is true

m = approximate number of preferences required when the alternative hypothesis is true

r = number of observations in the case of a fixed sample

Example 4.3: Suppose a sequential plan is desired for $\alpha = \beta = .05$ to test the hypothesis:

$$H: \quad \theta = 0.5$$
$$A: \quad \theta_0 = 0.8, \ \theta_1 = 0.2$$

Then the upper and lower boundaries of the rejection region are given by:

$$\text{upper:} \quad y = 4.29 + 0.3219n$$
$$\text{lower:} \quad y = -4.29 - 0.3219n$$

Similarly, the boundaries of the acceptance regions are given by:

$$\text{upper:} \quad y = -5.25 + 0.3219n$$
$$\text{lower:} \quad y = 5.25 - 0.3219n$$

The boundaries are graphed in Figure 4.2.

(ii) *Continuous case.* If the response is a continuous random variable that is normally distributed, we have similar sequential plans. The preference of treatment A over B is now given in terms of the difference of the measurements for treatments A and B. Let this be denoted by random variable Z. If treatments A and B are equivalent, the mean of Z is zero; otherwise we assume it to be δ or $-\delta$, according to whether A is preferred to B or B is preferred to A. We assume that the variance of the normal distribution is known and is taken to be equal to one for simplicity. That is, we test the hypothesis:

$$H: \quad \mu = 0$$
$$A: \quad \mu = \delta \quad \text{or} \quad \mu = -\delta$$

The stopping rule is given in terms of $y = \sum_{i=1}^{n} Z_i$ at any stage n. The

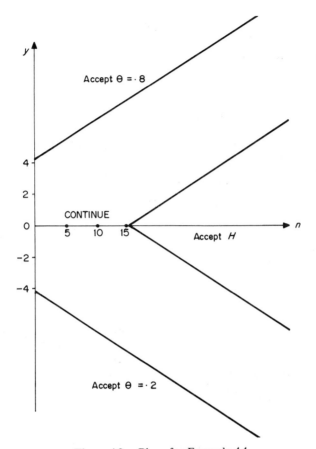

Figure 4.2. Plans for Example 4.1.

boundaries are again in terms of straight lines. A few cases are shown in Table 4.4, where a_1, a_2, b,... have the same meaning as in the case of binomial sequential plans.

Example 4.4: Trials to detect the difference between the use of calcium chloride (A) and adrenaline (B) as a bronchial dilator were made by Kilpatrick

Table 4.4   Open plan for $\alpha = .05 = \beta$

δ	a_1	a_2	b	l	m	r
0.2	18.19	14.85	0.1	205	165	325
0.4	9.09	7.43	0.2	51	42	82
0.8	4.55	3.71	0.4	13	10	21
1.4	2.60	2.12	0.7	4	3	7

and Oldham (1954). Each subject received the two substances to inhale in a random order. The rate of flow (expiration) was measured before and after the inhalation of each substance and the gain was measured. The hypotheses tested are

$$H: \quad \mu = 0$$

$$A: \quad \mu = .4 \quad \text{or} \quad \mu = -.4$$

with $\alpha = \beta = .05$. The boundaries of the acceptance region are given in Table 4.4.

$$\text{upper:} \quad y = -9.09 + .2n$$

$$\text{lower:} \quad y = 9.09 - .2n$$

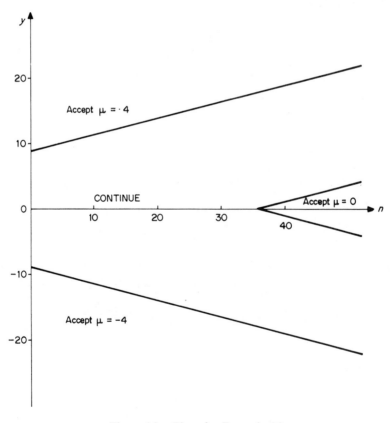

Figure 4.3. Plans for Example 4.2.

Similarly, the boundaries of the rejection region are given:

$$\text{upper:} \quad y = 7.43 + .2n$$

$$\text{lower:} \quad y = -7.43 + .2n$$

These boundaries are shown in Figure 4.3

When the variance is unknown, we would have a sequential t-test.

Exercises

6. For binomial sequential plans, find the boundaries of the rejection and acceptance regions from Table 4.3 and plot them for testing the hypothesis:

   (i) $H: \quad \theta = .5$

        $A: \quad \theta = .6, \quad \theta = .4$

   (ii) $H: \quad \theta = .5$

        $A: \quad \theta = .9, \quad \theta = .1$

   Assume $\alpha = \beta = .05$.

7. A blood-pressure-reducing drug (A) is to be tested clinically and compared with the standard drug (B). Assume that the numbers have been standardized so that the variance is one and the blood pressures are assumed to be normally distributed. Give the boundaries of the sequential procedure for testing the hypothesis:

   $$H: \quad \mu = 0$$
   $$A: \quad \mu = .2 \quad \text{or} \quad \mu = -.2$$

   Assume $\alpha = \beta = .05$. Graph the boundaries.

Chapter Exercises

1. A clinical trial involving 100 patients with terminal cancer of various kinds and a control group of 1,000 patients, untreated and matched, was made by Cameron and Pauling (1978). The treatment consisted of 10 g/day of vitamin C by intravenous infusion for about 10 days and orally thereafter. The treatment was discontinued if in the opinion of at least two independent clinicians the continuance of any conventional treatment offered no further benefit. For each treated patient, 10 control patients of the same sex and within 5 years of the same age were found who had suffered from cancer of the same primary organ and histological type. For illustration, the results on breast cancer (11 females) are given.

A = Survival time in days of an ascorbic acid (vitamin C) treated patient
B = Mean survival time of the 10 matched-control patients.

The clinical study was not double-blind.

Age	56	57	53	68	68	53
A	1235	24	1581	1166	40	727
B	796	977	1623	555	1304	1165

Age	75	68	55	43	53
A	3808	791	1804	3460 +	719
B	675	871	916	1311	978

Compare the two survival curves.

Summary

Clinical trials are field experiments conducted on human subjects. A good trial requires that it be *double-blind* so that the patient as well as the physician does not know the identity of the experimental intervention (drug or treatment). The control group in a clinical trial is usually given a *placebo*. Usual experimental designs such as *randomized blocks* and *Latin squares* can be used in clinical trials. *Historical controls* are also used in trials where the diseases are terminal. *Crossover designs* are usually used in the same subject when treatments are given in various orders and assumptions of no carry-over effects are made. *Sample sizes* can be determined in clinical trials when the significance level, power, and smallest detectable differences are given. *Sequential clinical trials* can be used to minimize the number of subjects on trial, if we can use the result of the subject for decision making.

References

Armitage, Peter. *Sequential Medical Trials*, Springfield, Ill.: Charles C. Thomas, 1960.
———. Exclusions, losses to follow-up, and withdrawals in clinical trials, in *Clinical Trials:Issues and Approaches*, Shapiro, S. H., and Louis, T. A., eds. New York: Marcel Dekker, 1983. pp. 99–114.
———, and Gehan, E. A. Statistical methods for the identification and use of prognostic factors, *Int. J. Cancer, 1974, 13*, 16–36.
Beta-Blocker Heart Attack Trial Research Group: A randomized trial of propranolol in patients with myocardial infarction. 1. Mortality Results. *J. Am. Med. Assoc.*, 1982, *242*, 1707–14.
Cameron, E., and Pauling, L. Supplemental ascorbate in the supportive treatment of cancer: Reevaluation of prolongation of survival times in terminal human cancer. *Proc. Natl. Acad. Sci.*, 1978, *75*, 4538–42.

Canner, Paul L. Practical aspects of decision making in clinical trials: The Coronary Drug Project as a case study, *Controlled Clinical Trials*, 1981, *1*, 363–76.

Elandt-Johnson, Regina C., and Johnson, Norman L. *Survival Models and Data Analysis*, New York: John Wiley & Sons, 1980.

Friedman, Lawrence M.; Furberg, Curt D.; and DeMets, David L. *Fundamentals of Clinical Trials*, Littleton, Mass: John Wright PSG, 1982.

Gould, A. Lawrence, and Pecore, Victor J. Group sequential methods for clinical trials allowing early acceptance of H_0 and incorporating costs, *Biometrika*, 1982, *69*, 75–80.

Heady, J. A. A cooperative trial in the primary prevention of Ischoemic heart disease using Clofibrate: Some statistical aspects, *Controlled Clinical Trials*, 1981, *1*, 383–92.

Hill, A. B., ed. *Controlled Clinical Trials*, Oxford: Blackwell, 1960.

Lachin, John M. Introduction to sample size determination and power analysis for clinical trials, *Controlled Clinical Trials*, 1981, *2*, 93–113.

Lebaeqz, Karen. Ethical aspects of clinical trials, in *Clinical Trials: Issues and Approaches*, Shapiro, S. H., and Louis, T. A., eds. New York: Marcel Dekker, 1983. Pp. 81–98.

Meier, Paul. Statistical Analysis of Clinical Trials, in *Clinical Trials*, Shapiro, S. H., and Louis, T. A. eds. New York: Marcel Dekker, 1983. Pp. 155–89.

Peto, R.; Pike, M. C.; Armitage, P.; Breslow, N. E.; Cox, D. R.; Howard, S. V.; Mantel, N.; McPherson, K.; Peto, J.; and Smith, P. G. Design and analysis of randomized clinical trials, requiring prolonged observation of each patient. II. Analysis & Examples, *Br. J. Cancer*, 1977, *35*, 1–39.

Pocock, S. J. Group sequential methods in design and analysis of clinical trials, *Biometrika*, 1977, *64*, 191–99.

Shapiro, S. H., and Louis, T. A., eds. *Clinical Trials: Issues and Approaches*, New York: Marcel Dekker, 1983.

Wilson, Gregory A.; McDonald, Clement J.; and McCabe, George P., Jr. The effect of immediate access to a computerized medical record on physician test ordering: A controlled clinical trial in the emergency room, *Am. J. Public Health*, 1982, *72*, 698–702.

Zelen, M. Play the winner rule and the controlled clinical trial, *J. Am. Statist. Assoc.*, 1969, *64*, 131–46.

———. The randomization and stratification of clinical trials, *J. of Chronics Diseases*, 1974, *27*. 365–75.

———. A new design for randomized clinical trial, *N. Engl. J. Med.*, 1979, *300*, 1241–45.

chapter five

Quality Assurance

Quality assurance is concerned with techniques to ascertain that the quality of a product conforms to established standards. Techniques of statistical quality control are routinely used in manufacturing and production in industry. The success of quality-control methods in Japanese industry is well recognized. Sampling inspection is another area of quality assurance and is used by large-scale consumers such as the U.S. Defencee Department and major corporations such as General Motors who order components from smaller vendors.

Another area of quality assurance is "reliability." When a product has many components, the theory of reliability is needed to provide answers to various questions about its successful performance.

The statistical study of quality assurance was initiated by Walter Shewhart of Bell Laboratories in 1924, when he developed "control charts" for use in production. The charts were able to point out the instance at which the production went out of control. Sampling inspection and reliability theory found their applications in World War II, when large-scale systems were under development and in production. The modern growth of high technology has focused even more attention to quality assurance. Military standards by the U.S. Defense Department were developed for acceptance sampling for inspection. These standards are used even today by government and industry for acceptance sampling and are based on statistical methods.

Whether we are estimating the life of an electric appliance or studying the performance of an airplane engine, we are using concepts of the theory of reliability. Quality-control methods are used not only by manufacturers but also by clinical chemists, radiologists, and endocrinologists. See McLemore (1981), Whitehead (1977), and Wilson et al. (1979).

This chapter discusses the three areas of quality assurance: quality control, sampling inspection, and reliability. In all of these areas, the fundamental object is to provide full satisfaction for the customer, a goal of every manufacturer or supplier.

The literature is extensive in the field of quality assurance. A few representative publications are listed at the end of the chapter. The *Journal of Quality Technology* is completely devoted to modern methods in quality assurance.

Quality-Control Charts

Control charts provide a graphical technique of testing a hypothesis about the parameter of a population. Assume that the measured characteristic of product quality is a random variable distributed normally with mean μ and variance σ^2. Then the probability that an observation is between three sigmas of the mean is .9973. That is, $P(\mu - 3\sigma < X < \mu + 3\sigma) = .9973$. A control chart for an observation has an upper control limit of $\mu + 3\sigma$ and a lower control limit of $\mu - 3\sigma$. These limits are drawn on a graph. Then the values of the random variable are plotted on this graph to see if they lie between the control limits. When the chart is made for measurements of quality characteristic of an item, any value that falls outside the control limits shows that the process needs to be checked. If the process producing the items is stable, the values should be within the control limits with very high probability. We consider several kinds of control charts in this section. Sometimes control charts use two-sigma limits. This implies that the probability is .9564 under the assumption of normality that an observation is within two standard deviations of the mean. All the examples here have three-sigma limits.

Control Chart for $\bar{x}$ (standards known)

Suppose that the quality characteristic is a continuous random variable with known mean μ and known standard deviation σ. The manufacturing process is said to be in *statistical control* if the items produced *conform* to the above specification. To provide a check, control limits for the mean of a sample of items randomly selected from the process are given. The *upper control limit* (UCL) is given in terms of mean and variance of $\bar{X}$. We know that $E(\bar{X}) = \mu$ and $\text{Var}(\bar{X}) = \sigma^2/n$. Hence the upper control limit is $\mu + 3\sigma/\sqrt{n}$. Similarly, the *lower control limit* (LCL) is $\mu - 3\sigma/\sqrt{n}$. μ is said to be the central line (CL). We then have:

$$\text{UCL} = \mu + 3\sigma/\sqrt{n} = \mu + A\sigma$$
$$\text{CL} = \mu \qquad\qquad = \mu$$
$$\text{LCL} = \mu - 3\sigma/\sqrt{n} = \mu - A\sigma$$

The constant $A = \dfrac{3}{\sqrt{n}}$ and its values are given in Table 5.1.

Table 5.1 Factors for Computing Control Limits

Number of observations in sample, n	$\bar{X}$ chart Factors for control limits		R chart Factor for central line	R chart Factors for control limits		s chart Factor for central line	s chart Factors for control limits		$\hat{\sigma}$ chart (biased) Factor for central line	$\hat{\sigma}$ chart (biased) Factors for control limits	
	A	A_2	d_2	D_3	D_4	c'_2	B'_2	B'_4	c_2	B_2	B_4
2	2.121	1.880	1.128	0	3.267	0.798	0	2.298	0.5642	0	3.267
3	1.732	1.023	1.693	0	2.575	0.886	0	2.111	0.7236	0	2.568
4	1.500	0.729	2.059	0	2.282	0.921	0	1.982	0.7979	0	2.266
5	1.342	0.577	2.326	0	2.115	0.940	0	1.889	0.8407	0	2.089
6	1.225	0.483	2.534	0	2.004	0.951	0.085	1.817	0.8686	0.030	1.970
7	1.134	0.419	2.704	0.076	1.924	0.960	0.158	1.762	0.8882	0.118	1.882
8	1.061	0.373	2.847	0.136	1.864	0.965	0.215	1.715	0.9027	0.185	1.815
9	1.000	0.337	2.970	0.184	1.816	0.969	0.262	1.676	0.9139	0.239	1.761
10	0.949	0.308	3.078	0.223	1.777	0.973	0.302	1.644	0.9227	0.284	1.716
11	0.905	0.285	3.173	0.256	1.744	0.976	0.336	1.616	0.9300	0.321	1.679
12	0.866	0.266	3.258	0.284	1.716	0.977	0.365	1.589	0.9359	0.354	1.646
13	0.832	0.249	3.336	0.308	1.692	0.980	0.392	1.568	0.9410	0.382	1.618
14	0.802	0.235	3.407	0.329	1.671	0.981	0.414	1.548	0.9453	0.406	1.594
15	0.775	0.223	3.472	0.348	1.652	0.982	0.434	1.530	0.9490	0.428	1.572
16	0.750	0.212	3.532	0.364	1.636	0.984	0.454	1.514	0.9523	0.448	1.552
17	0.728	0.203	3.588	0.379	1.621	0.984	0.469	1.499	0.9551	0.466	1.534
18	0.707	0.194	3.640	0.392	1.608	0.986	0.486	1.486	0.9576	0.482	1.518
19	0.688	0.187	3.689	0.404	1.596	0.986	0.500	1.472	0.9599	0.497	1.503
20	0.671	0.180	3.735	0.414	1.586	0.987	0.513	1.461	0.9619	0.510	1.490
21	0.655	0.173	3.778	0.425	1.575	0.988	0.525	1.451	0.9638	0.523	1.477
22	0.640	0.167	3.819	0.434	1.566	0.988	0.536	1.440	0.9655	0.534	1.466
23	0.626	0.162	3.858	0.443	1.557	0.989	0.546	1.432	0.9670	0.545	1.455
24	0.612	0.157	3.895	0.452	1.548	0.989	0.556	1.422	0.9684	0.555	1.445
25	0.600	0.153	3.931	0.459	1.541	0.990	0.566	1.414	0.9696	0.565	1.435

Reprinted with permission of ASTM, 1916 Race Street, Philadelphia, PA 19103.

The limits are drawn on a graph and the sample averages are plotted on this graph. The process is in control so long as the averages are within the limits. When the average is outside the limits, the process is said to be *out of control*. The process is stopped and checked before manufacturing is continued.

Example 5.1: The stiffness of aluminum alloy manufactured is assumed to have mean 2400 and standard deviation 80 in some standard units. Five were tested for stiffness daily with the following averages:

Day	1	2	3	4	5	6	7
Stiffness average	2450	2350	2505	2450	2375	2475	2550

The limits are given below. From Table 5.1, for $n = 5$, for $n = 5$, $A = 1.342$, so that:

$$UCL = 2400 + 1.342(80) = 2507$$
$$CL = 2400$$
$$LCL = 2400 - 1.342(80) = 2293$$

The limits and averages are shown on the control chart in Figure 5.1. We find that the process is out of control on day 7.

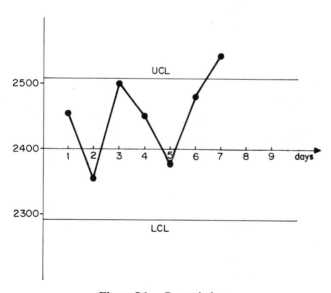

Figure 5.1. Control charts.

Control Charts for $\bar{X}$ (standards unknown)

Suppose the mean of the process-quality characteristic is not known a priori and must be estimated from the production itself. Based on k samples of size n each, we then have the overall average, where $\bar{X}_i$ is average of the ith sample:

$$\bar{\bar{X}} = \sum_{i=1}^{k} \bar{X}_i$$

Also, the overall range, where R_i is the range of the ith sample, is:

$$\bar{R} = \sum_{i=1}^{k} R_i$$

The charts are given by:

$$UCL = \bar{\bar{X}} + A_2 \bar{R}$$
$$CL = \bar{\bar{X}}$$
$$LCL = \bar{\bar{X}} - A_2 \bar{R}$$

The constants A_2 are tabulated in Table 5.1. Notice that A_2 is based on the sample range whereas A was based on known σ.

Control charts for Range R

The control limits for the sample range based on the sample size n are given below in terms of $\bar{R}$:

$$UCL = D_4 \bar{R}$$
$$CL = \bar{R}$$
$$LCL = D_3 \bar{R}$$

The constants D_3 and D_4 are obtained so as to provide the control limits and are tabulated in Table 5.1.

If the population standard deviation σ is known, the limits for the sample range can be obtained from:

$$UCL = D_2 \sigma$$
$$CL = d_2 \sigma$$
$$LCL = D_1 \sigma$$

where $D_1 = d_2 D_3$ and $D_2 = d_2 D_4$. Limits can be obtained from tabulated values of d_2, D_3 and D_4 from Table 5.1.

Control Chart for Proportion-Defectives

Suppose the manufacturing process is producing items with the proportion defectives (nonconforming items), p. Then the three-sigma limits are given by:

$$UCL = p + 3\sqrt{\frac{p(1-p)}{n}}$$

$$CL = p$$

$$LCL = p - 3\sqrt{\frac{p(1-p)}{n}}$$

The sampling scheme provides samples of size n, and the estimate of p is obtained by the ratio of the total number of defectives in the sample and the sample size. These are plotted and point out if the process is out of control. Note that the proportion-defectives, p, is always greater than or equal to zero. Hence, if the lower control limit turns out to be negative, the effective limit is zero.

Example 5.2: Suppose the proportion defectives of a process are $p = .3$. Samples of size 100 are taken with the following results:

Sample number	1	2	3	4	5	6
Number of defectives	4	2	1	5	2	3
Percent defective	.04	.02	.01	.05	.02	.03

The control limits are:

$$UCL = .03 + \sqrt{(.03)(.97)/100} = .047$$

$$CL = .03$$

$$LCL = .03 - \sqrt{(.03)(.97)/100} = .013$$

The control chart is given in Figure 5.2. The figure shows that between sample numbers 3 & 4 the process goes out of control and should be checked for assignable causes.

Variable Group Size in p-Chart

When the sample sizes of various sampled groups are unequal, we have the control limits:

$$UCL = \bar{p} + 3\sqrt{\bar{p}(1-\bar{p})}$$

$$CL = \bar{p}$$

$$LCL = \bar{p} - 3\sqrt{\bar{p}(1-\bar{p})}$$

where

$$\bar{p} = \frac{\text{Total defectives from various samples}}{\text{Total sample size}}$$

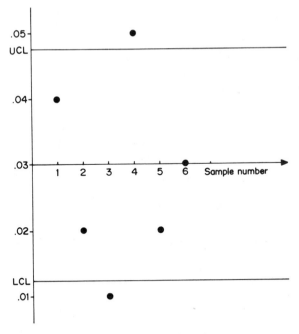

Figure 5.2. Control chart for p.

Control Chart for the Total Number of Defectives

The chart for the total number of defects can be used in place of the p-chart if the number of sample observations is the same from group to group. Here the following limits arise:

$$\text{UCL} = np + 3\sqrt{np(1-p)}$$
$$\text{CL} = np$$
$$\text{LCL} = np - 3\sqrt{np(1-p)}$$

c-chart

When the number of observations sampled is large and the proportion of defectives is small, the number of defective items has a Poisson distribution. In many industrial situations, the Poisson distribution arises naturally. For example, the surface imperfections in a painted area or the number of defects in a piece of cloth have a Poisson distribution.

Since the standard deviation of a Poisson random variable is the square root of the mean, the three-sigma limits are:

$$UCL = c + 3\sqrt{c}$$
$$CL = c$$
$$LCL = c - 3\sqrt{c}$$

Negative control limits are usually ignored and only the upper control limit is used. When each observation is based on a sample of size n, the control limits are:

$$UCL = c + 3\sqrt{\frac{c}{n}}$$
$$CL = c$$
$$LCL = c - 3\sqrt{\frac{c}{n}}$$

Example 5.3: It has been determined that on the average, 3 defects per square yard of a piece of cloth are found in a factory. A sample of 5 square yards is taken every day to check the defects. Assuming that the number of defects has a Poisson distribution, we have the limits:

$$UCL = 3 + 3\sqrt{3/5} = 5$$
$$CL = 3$$
$$LCL = 3 - 3\sqrt{3/5} = 1$$

Control Limits for Standard Deviation

Suppose k samples of size $n_1, \ldots, n_k$ are given with estimated sample variances:

$$s_i^2 = \frac{\sum(X_j - \bar{X})^2}{n_i - 1}$$

Then the pooled sample variance is:

$$s_p^2 = \sum_{i=1}^{k} (n_i - 1)s_i^2 / \sum(n_i - 1)$$

The control limits of a sample standard deviation based on sample size n are:

$$UCL = B'_4 s_p$$
$$CL = c'_2 s_p$$
$$LCL = B'_2 s_p$$

If we use the biased estimate of the standard deviation $\hat{\sigma}$. we have:

$$UCL = B_4 \hat{\sigma}$$
$$CL = c_2 \hat{\sigma}$$
$$LCL = B_2 \hat{\sigma}$$

The above constants are given in Table 5.1.

Exercises

1. In a manufacturing line, the standard is given in terms of mean 25 and standard deviation 5. Find the control limits for the mean and plot the following data to see if the process goes out of control when each sample is based on 10 observations:

Sample 1	1	2	3	4	5	6	7	8
Average	22	25	27	23	27	26	33	29

2. Give the control limits for the $\bar{X}$ and R chart using the following data, $n = 5$:

Sample	1	2	3	4	5	6	7	8	9	10
$\bar{X}$	201	199	204	197	198	202	200	202	198	195
R	13	20	7	16	8	9	10	16	7	20

3. What are the control limits for the proportion of defectives in samples of 50 items selected every hour from a production line so that the standard fraction of defectives is kept at .01? Given the following data, plot it on the control chart to see if the process is in statistical control:

Sample	1	2	3	4	5	6	7	8	9	10
Number of defectives	1	0	0	1	2	0	2	1	0	0

4. In Exercise 3, give control limits for the total number of defectives (that is, charts for np).
5. The production of textiles in a factory is beging monitored for defects. It is generally assumed that 3 defects per square yard are standard. Find the control limits using the c-chart for defects per square yard.

Sampling Inspection by Attributes

An item may be regarded as defective (nonconforming) or nondefective (conforming). *Sampling inspection by attributes* can be completed in one of two ways:

(i) *Lot-by-lot inspection.* The material to be inspected is divided among large lots containing a known number of items.

(ii) *Continuous inspection.* The material is produced on a machine continuously and has to be inspected by sampling items from the continuous production. For information on continuous inspection, see Schilling (1982).

When a certain characteristic of the item is being measured, we have *sampling inspection by variable*, which is considered in the next section.

In attribute-sampling inspections, we test hypotheses about the proportion-defectives. If the proportion-defectives is below a certain number, we accept the lot; otherwise we reject the lot. Two kinds of errors arise in testing a hypothesis. In sampling inspection, similar errors arise.

We may reject a good lot or accept a bad lot. The probabilities of these errors are known as *producer's risk* and *consumer's risk*. The probability of rejecting a good lot is called the *producer's risk*. It corresponds to α, the probability of a type I error in testing hypothesis. The probability of accepting a bad lot is called the *consumer's risk*. It corresponds to β, the probability of a type II error in testing hypotheses.

Another important quantity that is used in measuring the performance of a sampling plan is the *operating characteristic* curve. This curve gives the probability of accepting the lot for various values of the proportion-defectives in sampling by attributes. If p is the proportion-defectives in a lot, then the probability of a type II error, $\beta(p)$, is the operating characteristic function, or *OC function*. OC function is one minus the power function of the test about p.

A typical OC curve is given in Figure 5.2. The probability of defectives where the producer's risk is α (the value of the Operating Characteristic curve is $1 - \alpha$) is called the *producer's quality level* (PQL), or *acceptable quality level* (AQL), denoted by p_1. Similarly, the value where the OC curve attains consumer's risk is called *consumer's quality level* (CQL) and is usually denoted by p_2. It is also known as the *lot tolerance percent defective* (LTPD).

P_1, the proportion-defectives a producer can allow, is usually small, since a producer wants good lots to be accepted and the consumer wants bad lots to be rejected. Figure 5.3 shows the PQL, CQL, and producer's and consumer's risk.

We shall consider *single sampling* and *double sampling* plans for attributes.

Single sampling plans: Given a sample size n and a number c, the plan specifies to reject the lot if the number of defectives in the sample is larger than c. That is, we have the following rule in terms of X, the number of defectives in a sample of size n.

$$\text{Accept the lot if } X \leq c$$
$$\text{Reject the lot if } X > c$$

Double sampling plans: The sample is taken in two steps. First, a random sample of size n_1 is taken and two numbers c_1 and c_2 are given. The rule

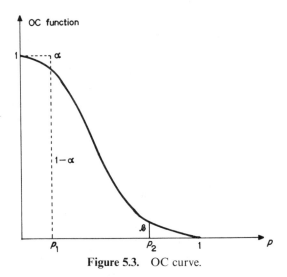

Figure 5.3. OC curve.

specifies the following. Let X be the number of defectives in the first sample. Then:

$$\text{Accept the lot if } X < c_1$$
$$\text{Reject the lot if } X > c_2$$

Take another sample if $c_1 < X < c_2$. The size of the second sample is n_2. Let Y be the number of defectives in the second sample. Then:

$$\text{Reject the lot if } X + Y > c_3$$
$$\text{Accept the lot if } X + Y \le c_3$$

Extensive tables by Dodge and Roming (1959) provide the constants in single and double sampling plans for various risks. Double sampling plans were a precursor of sequential plans, which will be discussed later.

Average outgoing quality and average outgoing quality limit for single sampling plans: Let p be proportion defectives in a lot of size N. Then we expect Np defectives in the lot. A random sample of size n is taken without replacement and let X be the number of defectives in the sample. Then the probability that $X = x$ is given by a hypergeometric distribution, with density $f(x|p)$. The expression for this density is given in Chapter 3, Volume I. For single sampling plans, we accept the lot if $X \le c$. Hence the probability of accepting the lot is:

$$P(A|p) = \sum_{x=0}^{c} f(x|p)$$

$P(A|p)$ is the *operating characteristic* of the above sampling plan.

Average outgoing quality (AOQ): When the proportion-defectives is p, the average outgoing quality is:

$$AOQ = p \cdot P(A|p)\frac{N-n}{N}$$

The lot is accepted if $p = 0$; hence $P(A|0) = 1$ and $P(A|1) = 0$ when $p = 1$. The AOQ curve starts at zero and ends at zero. For some value of p, AOQ attains a maximum. The maximum value of AOQ is called the average outgoing quality limit (AOQL).

Example 5.4: Suppose a sample of 3 is chosen without replacement from a lot of size 10. The lot is accepted if the number of defectives in the sample is zero. That is, we have $N = 10$, $n = 3$, $c = 0$. The probability of accepting the lot when the percent defective is p is given by:

$$P(A|p) = \binom{10-10p}{3}\binom{10p}{0} \bigg/ \binom{10}{3}$$

$$= \binom{10-10p}{3} \bigg/ \binom{10}{3}$$

Hence,

$$AOQ = .7p\binom{10-10p}{3} \bigg/ \binom{10}{3}$$

So we have:

p	0	.1	.2	.3	.4	.5	.6	.7	.8
AOQ	0	.049	.065	.061	.047	.029	.014	.004	0

This table shows that AOQL $= .065$. That is, the maximum proportion-defectives in the accepted lots will be .065.

An extensive listing of AOQs and AOQLs for various sampling plans is available in Dodge and Roming (1959).

Exercises

6. Find the AOQ for the following sampling inspection schemes (sampling without replacement):

$$\text{(i) } N = 10, \quad n = 4, \quad c = 0$$
$$\text{(ii) } N = 15, \quad n = 2, \quad c = 0$$
$$\text{(iii) } N = 25, \quad n = 2, \quad c = 1$$

7. When sampling is done with replacement, the binomial distribution can be used in place of the hypergeometric distribution. Solve Exercises 6 in this case.

8. For lots of size 7, consider a single sampling plan with $n = 4$ and $c = 0$. What is the probability of acceptance of the lot if $p = .125$ and $p = .50$ (sampling without replacement)?

Sampling Inspection by Variables

In some cases, the quality characteristic can be described by a continuous random variable such as the tensile strength of a wire, the weight of a box of cereal, the length of a screw and so on. We assume now that the quality characteristic is measured by a random variable X, which is normally distributed with mean μ and variance σ^2. The sampling inspection will accept the lot if $L \le X \le U$ and reject the lot if $X > U$ or $X < L$, where L and U are the lower and upper specification limits. In place of the double specification limits, we may have single specification limits. That is, we reject the lot if $X < L$; otherwise we accept it. Similarly, we reject the lot if $X > U$; otherwise we accept it. To determine single specification limits, suppose now that the producer's risk is α and the consumer's risk is β. Let $PQL = p_1$ and $CQL = p_2$. Consider the following two cases.

(1) L is given. Let $\bar{X}$ be the sample average. The rule is:

$$\text{Accept the lot if } \bar{X} - k\sigma \ge L$$
$$\text{Reject the lot if } \bar{X} - k\sigma < L$$

The constant k can be determined by the equation:

$$k = \frac{z_{1-p_2} z_{1-\alpha} + z_{1-p_1} z_{1-\beta}}{z_{1-\alpha} + z_{1-\beta}}$$

Here $z_{1-\theta}$ is the $100(1 - \theta)$th percentile of the standard normal distribution.

(2) U is given. The rule is:

$$\text{Accept the lot if } \bar{X} + k\sigma \le U$$
$$\text{Reject the lot if } \bar{X} + k\sigma > U$$

The constant k is the same as given above. The sample size can be obtained as:

$$n = \left(\frac{z_{1-\alpha} + z_{1-\beta}}{z_{1-p_1} - z_{1-p_2}} \right)^2 n$$

When σ is not known, a modification of the above sample size is given by n_0, where

$$n_0 = \left(1 + \frac{k^2}{2} \right)$$

When both specification limits are to be used, the procedures are complicated. For reference, see Schilling (1982).

Example 5.5: The resistance of an electrical device is to be monitored. $p_1 = .05$, $p_2 = .10$, $\alpha = .05$, and $\beta = .10$. So we have:

$$k = \frac{(1.28)(1.645) + (1.645)(1.28)}{1.645 + 1.28} = 1.44$$

The sample size when σ is known is:

$$n = \left(\frac{1.645 + 1.28}{1.645 - 1.28}\right)^2 = 64$$

If σ is not known, $n_0 = \left(1 + \frac{1.44^2}{2}\right)(64) = 130.$

Tolerance Intervals

The specifiations for upper and lower limits are often given in terms of a *tolerance interval*. Tolerance intervals provide the limits that contain a specified proportion of a population with a given confidence level. For example, a tolerance interval (a, b) may contain 90% of the population with probability 95%. Then 95% is the *confidence level* of the tolerance interval. Tolerance intervals should be distinguished from confidence intervals where the unknown parameter of the population is covered by the confidence interval with a given confidence level.

An example of a tolerance interval for 95% of a normal population with mean μ and variance σ^2 is $(\mu - 1.96\sigma, \mu + 1.96\sigma)$, and the confidence level here is 100%.

Exercises

9. Find the rule for acceptance of a lot when the lower specification limit is 75 and $p_1 = .01$, $p_2 = .10$, $\alpha = .05$, and $\beta = .10$ in terms of $\bar{X}$ given $\sigma = 10$.
10. Find the rule for the acceptance of a lot when the upper specification limit is given as 90 in Exercise 9.
11. The producer's quality level is .05 and the consumer's quality level is .20. Find k and n when $\alpha = \beta = .05$ in sampling inspection by variables.

Sequential Sampling Inspection

In sequential sampling we have to accept or reject the hypothesis (or the lot) or take an additional observation at each stage of sampling. Some aspects of sequential experimentation as applied to medical trials were given in Chapter 4. This section considers sequential plans for attribute sampling and variable sampling.

Attribute Sampling Plan

Suppose the acceptable quality level (AQL) in a lot is given by p_1 and the rejectable quality level (RQL) is given by p_2. That is, if the proportion-defectives is p_1, we accept the lot; if it is p_2, we reject the lot. We are then testing the null hypothesis, $p = p_1$, versus the alternative, $p = p_2$.

The observations are taken in a sequence. At stage n, we have n observations. Let d be the number of defectives in the sample. The sequential rule is given in terms of constants A and B with $A = \dfrac{1 - \beta}{\alpha}$ and $B = \dfrac{\beta}{1 - \alpha}$, where α and β are producer's and consumer's risks, respectively.

Decision Rule:

(i) Accept H if $d \leq B$
(ii) Reject H if $d \geq A$
(iii) Continue sampling if $B < d < A$

The procedure can be graphed with the help of parallel lines,

$$d = -h_1 + gn$$
$$d = h_2 + gn$$

where $h_1 = b/c$, $h_2 = a/c$ and

$$a = \log A, \quad b = -\log B$$

$$c = \log \frac{p_2(1 - p_1)}{p_1(1 - p_2)}$$

$$g = \left(\log \frac{1 - p_1}{1 - p_2} \right) \Big/ c$$

Typical regions are described in Figure 5.4. These lines divide the rejection, acceptance, and continuation regions. The sequential sampling plan can now be implemented with the help of this graph. At any stage, we plot (n, d) and decide upon the appropiate action depending upon the region where this point is located.

Example 5.6: Suppose we want a sequential sampling plan for:

$$p_1 = .01, \quad p_2 = .08, \quad \alpha = .05, \quad \beta = .10$$

We find:

$$A = \frac{1 - .10}{.05} = 18, \quad B = \frac{.10}{1 - .05} = .105$$

$$a = \log A = 1.255, \quad b = -\log B = .978$$

$$c = \log \frac{(.08)(.99)}{(.01)(.92)} = .935$$

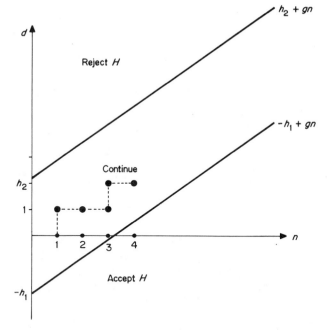

Figure 5.4. Sequential sampling plans.

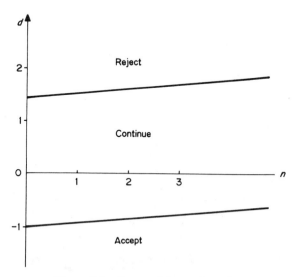

Figure 5.5. Sequential sampling plans.

$$h_1 = .978/.935 = 1.046$$
$$h_2 = 1.255/.935 = 1.342$$
$$g = \log\frac{(.99)}{(.92)}/.935 = .034$$

The lines giving the boundaries of the acceptance, rejection, and continuation regions are:

$$d = -1.046 + .034n$$
$$d = 1.342 + .034n$$

The regions are graphed in Figure 5.5. The operating characteristic of the plan can also be obtained.

Variable sampling Plans

We assume that the measurement X on a sampling unit is normally distributed with mean μ and variance σ^2 (known). Suppose AQL is μ_1 and the RQL is μ_2 with $\mu_1 < \mu_2$. That is, we want to test the hypotheses, with consumer's and producer's risks given as before:

$$H: \quad \mu = \mu_1$$
$$A: \quad \mu = \mu_2$$

Then the rejection, acceptance, and continuation regions are given with the help of the lines

$$y = -h_1 + gn$$
$$y = h_2 + gn$$

where $y = \sum_{i=1}^{n} X_i$, $h_1 = b\sigma^2/(\mu_2 - \mu_1)$, $h_2 = a\sigma^2/(\mu_2 - \mu_1)$, $g = (\mu_1 + \mu_2)/2$, $a = \ln A$, and $b = -\ln B$.

Example 5.7: Let the mean $\mu_1 = 1.18$ and $\mu_2 = 1.19$, $\alpha = .10$, $\beta = .10$, and $\sigma = .0004$:

$$A = \frac{1 - .1}{.1} = 9 \quad B = \frac{.1}{1 - .1} = \frac{1}{9}$$
$$a = 2.197 \quad b = 2.197$$
$$h_1 = h_2 = (2.197)(.004)/(1.19 - 1.18)^2 = 8.788$$
$$g = \frac{1.19 + 1.18}{2} = 1.185$$

Hence the boundaries of the regions are:

$$y = -8.788 + 1.185n$$

and

$$y = 8.788 + 1.185n$$

Exercises

12. Give a sequential sampling plan for the number of defectives for the problem:

$$p_1 = .01. \quad p_2 = .03$$
$$\alpha = .05, \quad \beta = .1$$

13. Graph the rejection, acceptance, and continuation regions in Exercise 12.
14. For the following specifications, give the sequential sampling plan and graph the acceptance and rejection regions:

$$p_1 = .1, \quad p_2 = .3$$
$$\alpha = .10, \quad \beta = .15$$

15. In a variable-sampling situation, we are given that $\mu_1 = 2.1$, $\mu_2 = 2.5$, $\alpha = \beta = .1$, and $\sigma^2 = .04$. Find the sequential sampling plans.
16. Give the acceptance, rejection, and continuation regions for Exercise 15 on a graph.
17. For a process with variable inspection, we are given:

$$\mu_1 = 85, \quad \mu_2 = 90, \quad \sigma^2 = 25, \quad \mu = \beta = .1$$

Find the boundaries of the sequential sampling plan and sketch the regions on a graph.

System Reliability

Reliability is defined as "the probability of a device or an item performing its defined purpose adequately over a specified period of time, under available conditions." We would like to know the reliability of systems that have several components, when reliabilities of their components are known. A system is said to be a *series system* if the failure of a single component leads to the failure of the system. Diagrammatically, it means the components are joined in a series:

$$\rightarrow \boxed{1} \rightarrow \boxed{2} \rightarrow \boxed{3} \rightarrow \boxed{4} \rightarrow \boxed{5} \rightarrow$$

For example, a string of light bulbs joined by a single wire is a system in series. Suppose the reliability of the ith component is p_i, $i = 1, 2, \ldots, k$, for a system having k components in series. Assuming that the components are independent, the reliability of the system is:

$$R = p_1 p_2 \cdots p_k$$

If all components have the same reliability p, then:

$$R = p^k$$

For example, if a system of three components, each having a reliability of .9, is jointed in series, its reliability is $(.9)^3 = .729$.

A system is said to be *in parallel* if the system works when at least one component is working. Diagrammatically, a parallel system can be represented:

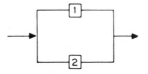

This system is said to have *redundant* components. When we have k independent components joined in parallel, and ith component has probability p_i, $i = 1, 2, \ldots, k$ of successful performance, the reliability of the system is given by:

$$R = 1 - (1 - p_1)(1 - p_2), \ldots, (1 - p_k)$$

If all components have the same reliability p, we have:

$$R = 1 - (1 - p)^k$$

For example, if the 3 components, each having a reliability of .9, are in a parallel system, the system reliability is:

$$R = 1 - (1 - .9)^3 = .999$$

Example 5.8:   For the guidance system of a rocket, three components were used in parallel, having a reliability of .92, .96, and .99, respectively. The reliability of the guidance system is:

$$R = 1 - (1 - .92)(1 - .96)(1 - .99)$$
$$= 1 - .000032 = .999968$$

Reliability Function

Let the lifetime of a device be random variable T. By the *reliability* in any given instant of time t, we mean the probability that the device is in operation at t. That is, the reliability at T is the probability that $T \geq t$. Let $F(t)$ be the cumulative distribution function of the lifetimes of the device. Then the reliability at time t is:

$$R(t) = 1 - F(t)$$

Suppose the lifetimes are distributed exponentially with parameter λ. That is:

$$F(t) = 1 - e^{-\lambda t}$$

The mean lifetime is $1/\lambda$ and the variance is $1/\lambda^2$ (see chapter 4), Volume I. Then the *reliability function* of the device is:

$$R(t) = 1 - (1 - e^{-\lambda t}) = e^{-\lambda t}$$

The reliability function is also known as the *survival function*. The techniques used in reliability theory are the same as in survival analysis (see Chapter 3).

For example, suppose we want to provide an estimate of the reliability of a device having exponential lifetimes. Let $t_1, t_2, \ldots, t_n$ be the n lifetimes of a random sample of n devices on test. The estimate of λ is given by:

$$\hat{\lambda} = \frac{1}{\bar{t}}$$

Hence the estimate of the reliability function is $\hat{R}(t) = e^{-t/\bar{t}}$.

Example 5.9: The lifetimes of electron tubes are exponentially distributed with parameter λ. Ten tubes provided an average lifetime of 1,500 hours. The reliability function is:

$$R(t) = e^{-\frac{t}{1500}}$$

The probability that a tube will have a lifetime of 1,650 hours is:

$$R(1650) = e^{-\frac{1650}{1500}} = e^{-1.1} = .33$$

Example 5.10: A system consists of four components with reliabilities, respectively, of .9, .8, .7, and .8 of components 1, 2, 3, and 4. The first two are connected in parallel and the last two in series, as shown below. Find the reliability of the system.

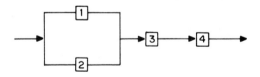

Assuming the independence of components, the reliability of the first two components is:

$$1 - (1 - .9)(1 - .8) = .98$$

Hence the reliability of the system is:

$$(.98)(.7)(.8) = .5488$$

Exercises

18. A system consists of 5 components in series. Their reliabilities are given as .7, .9, .8, .5, and .9. Find the reliability of the system.
19. Suppose a system consisting of three components is in parallel, with individual probabilities of successful performance given as .9, .7, and .8. Find the system reliability.
20. A system consists of five components, the first three in parallel and the last two in series. Their probabilities of successful performance are given as .8, .7, .9, .7, and .75, respectively. Find the system reliability.
21. Assume that the life of automobile tires has an exponential distribution with an expected life of 45,000 miles. What is the reliability function of the tires?
22. Given a random sample of 5 tires with lifetimes leading to an average value of 40,120 miles, find the estimated reliability function. What is the probability that a tire from this popultion will fail before 35,000 miles?

Chapter Exercises

1. The following data for a production line with a sample size of 5 is given. Find the three-sigma control limits for the $\bar{X}$, R, and s charts:

Sample number	1	2	3	4	5	6	7
$\bar{X}$	18.1	17.9	18.0	16.9	17.5	18.2	18.3
R	1.1	2.1	.9	.7	.8	1.2	1.4
s	.7	.8	1.1	.8	.9	.8	.7

2. Beer cans are inspected for visual defects and the sample size of each lot is 312 (see Burr, 1976, p. 145). Plot the data on a control chart for the number of defects (c-chart). Comment.

Lot number	1	2	3	4	5	6	7	8	9	10
Number of defectives	6	7	5	7	5	5	4	5	12	6

Lot number	11	12	13	14	15	16	17	18	19	20	21	22	23	24	25
Number of defectives	7	7	6	6	6	6	23	10	8	5	9	5	11	15	10

3. Consider a system with six components, with the first two and the last two connected in parallel, while the middle two connected in series. The reliability of the six components in order is .7, .85, .9, .8, .6, and .5. Find the reliability of the system.

Summary

Quality assurance is important in modern industry. *Control charts* provide a technique for checking if the process is in *statistical control*. There are several kinds of control charts. *Sampling inspection* or *acceptance sampling* is needed by large manufacturers and large government organizations such as the military. Inspection may be performed by *lot-by-lot* inspection or by *continuous inspection*. The acceptance rules may depend on the number of defective items in a sample, called *attribute sampling*, or some measurement of the item, called *variable sampling*. *Sequential sampling plans* require that at any stage either the lot is accepted or rejected or another observation is taken. The sequential sampling procedure can also be described on a graph.

Another important area of quality assurance is concerned with measuring the *reliability* of a *system* of components. The systems may be formed in *series* or *in parallel*. The *reliability function* describes the probability that a device will survive a given instant of time and is the same as the *survival function*. The simplest model used for lifetimes of devices is exponential distribution.

References

Burr, Irving, *Statistical Quality Control Methods*, New York: Marcel Dekker, 1976.
———. *Elementary Statistical Quality Control*, New York: Marcel Dekker, 1979.
Dodge, H. F., and Romig, H. G. *Sampling Inspection Tables: Single and Double Sampling*, 2d ed. New York: John Wiley & Sons, 1959.
Grant, Eugene L., and Leavenworth, Richard S. *Statistical Quality Control*, 4th ed. New York: McGraw-Hill, 1972.
Mann, Nancy R.; Schafer, Ray E.; and Singpurwalla, Nozer D. *Methods for Statistical Analysis of Reliability and Life Data*. New York: John Wiley & Sons, 1974.
McLemore, Joy M. *Quality Assurance in Diagnostic Radiology*. Chicago: Year Book Medical Publishers, 1981.
Ott, E. R. *Process Quality Control*, New York: McGraw-Hill, 1975.
Schilling, Edward G. *Acceptance Sampling in Quality Control*, New York: Marcel Dekker, 1982.
Whitehead, T. P. *Quality Control in Clinical Chemistry*, New York: John Wiley & Sons, 1977.
Wilson, D. W.; Gaskell, S. J.; and Kemp, K. W., ed. *Quality Control in Clinical Endocrinology*, Cardiff: Alpha Omega, 1979.

chapter six

Longitudinal Studies

The study of change of characteristics of a population is required in many contexts. In clinical trials, patient populations are followed over a long period to study the effectiveness of a treatment. To investigate the effect of diets on the growth of children, long-term studies are usually performed. In some cases, we are interested in observing a given event, such as failure of an electric bulb when the bulb is on for a long period. There are two ways in which the change of a characteristic of a population may be studied. One way is to take different groups of subjects at different instants of time and study the change, such studies are known as *cross-sectional* studies. The other method is to follow up the same subject for a long period of time. These studies are called *longitudinal*.

. Longitudinal studies may be retrospective or prospective. The measurements on a subject over a period of time are also known as a *time series*, but we will not consider it here. Longitudinal studies may also be performed where we are measuring the time until the occurrence of an event such as death. These kinds of studies are known are *incidence studies*. Examples of incidence studies have already occurred earlier in the form of survival studies and clinical trials. The other kind of longitudinal studies are *growth studies*, where continuous responses are measured on subjects over a period of time.

The basic questions in longitudinal studies are concerned with the comparison of populations, and studying relationships among various variables, at given instants of time or the relationship of variables at different times. This chapter presents examples of longitudinal studies and shows how they can be analyzed using methods of regression, correlation, and analysis of variance. References to several important recent longitudinal studies are given at the end of the chapter.

Examples of Longitudinal Studies

Several examples of incidence and growth studies.

Example 6.1: To measure the effect of antibiotics on mortality from infectious diseases, Hemminki and Paakulainen (1976) studied mortality rates for several diseases in Sweden and Finland. In this illustration, mortality rates of scarlet fever, tuberculosis, and acute gastroenteritis are used for the year 1911–66 in Sweden. In 1950, paraamino-salicylic acid (PAS) was introduced. The problem is to ascertain if mortality rates have changed since 1950. The data are recorded in Table 6.1.

Table 6.1   Mortality rates (per 100,000) for Sweden

Year	Scarlet Fever	Tuberculosis	Acute gastro-enteritis
1911	4.8	193.8	31.1
1916	5.6	181.0	22.2
1921	1.4	148.5	16.1
1926	1.1	131.5	12.1
1931	0.82	108.6	5.9
1936	1.6	81.5	4.2
1941	0.96	69.5	3.2
1946	0.12	39.6	2.2
1951	0.02	15.6	2.8
1956	0.00	8.3	3.0
1961	0.00	5.5	3.9
1966	0.02	4.4	3.7

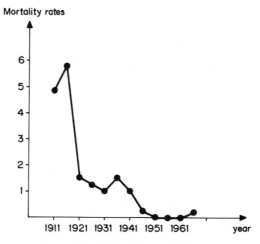

Figure 6.1.   Rates for scarlet fever.

The mortality rates were obtained by following the same population over the period 1911–66 at five-year intervals. They are plotted in Figure 6.1 for scarlet fever.

Figure 6.1 gives a growth-curve representation of data. Statistical techniques of regression, for example, may be needed to see whether the introduction of PAS in 1950 has any effect on mortality rates. One can use regression models for data before 1950 and after 1950 and compare them. We fit two linear regressions—one to data before 1950 and the other to data after 1950. The regression coefficients for tuberculosis mortality rates are

$$\hat{\beta}_1 = -4.4 \text{ for data before } 1950$$
$$\hat{\beta}_2 = -0.8 \text{ for data after } 1950$$

Similarly, for acute gastroenteritis, we have the slope of the regression lines given by

$$\hat{\beta}_1 = -1.1 \text{ for data before } 1950$$
$$\hat{\beta}_2 = -0.039 \text{ for data after } 1950$$

Example 6.2: In a study to find the relationship between age-specific marriage rates and legal abortions, Bauman, Anderson, Freeman, and Koch

---

Table 6.2   (a) Rank Correlations between Age-specific marriage rates & Legal Abortions for Example 6.2

			Ages 15–19 years		
Years	1966	1967	1968	1969	1970
1967	− 0.045				
1968	− 0.105	− .003			
1969	− 0.135	− .092	− .131		
1970	− 0.134	− .126	− .196	− .165	
1971	− 0.213	− .233*	− .297*	− .308*	− .252*

Table 6.2   (b)

			Ages 20–24 years		
Years	1966	1967	1968	1969	1970
1967	− .023				
1968	+ .012	− .013			
1969	− .082	.081	− .247*		
1970	− .105	.147	− .225*	.217	
1971	− .188	.233	− .341*	.336*	.278*

Table 6.2 (c)

Ages 25–29 years

Years	1966	1967	1968	1969	1970
1967	−.045				
1968	.158	.247*			
1969	.219	.328*	.140		
1970	.136	.209	.224	−.034	
1971	.039	.040	−.109	−.235*	.243*

Table 6.2 (d)

Ages 30–34 years

Years	1966	1967	1968	1969	1970
1967	− 0.220				
1968	− 0.139	−.020			
1969	− 0.111	.053	.054		
1970	− 0.018	.126	.139	.104	
1971	− 0.163	.046	−.070	−.051	−.096

Table 6.2 (e)

Ages 35–44 years

Years	1966	1967	1968	1969	1970
1967	−.112				
1968	−.059	.128			
1969	−.202	−.035	−.109		
1970	−.013	.070	.036	.143	
1971	−.169	−.023	−.120	.001	−.105

(1977) studied data from 38 states and the District of Columbia for the years 1966–71 for various age groups. The rank correlations were calculated between age-specific marriage rates and legal abortions based on the data from the states and are given in Table 6.2 for various age groups. The significant rank-correlation coefficients at level .05 are marked by an asterisk. The number of significant rank correlations are given by age in Table 6.3.

The absence of significant correlations in higher age groups indicates that

Table 6.3   Significant Correlations in Example 6.2.

Age	Number of significant Correlations
15–19	4
20–24	5
25–29	4
30–34	0
35–44	0

the relationship between marriage rates and legal abortions is stronger for younger age groups than older age groups. Hence, the hypothesis that legal abortions encourage postponement of marriage may be verified by the above argument.

In this example, the time series data have been used in a different way than in Example 6.1. We used the rank-correlation coefficients at various age groups to show that as time increases the number of significant correlations decreases. We may alternatively use a model of time series for marriage rates and legalized abortions. For example, we may assume that the marriage rates depend on the age and use a model in terms of age. Similarly, we can provide legalized abortions as functions of age and then use techniques to find correlations between the marriage raes and legalized abortions.

Exercises

1. Age-specific-level abortions per 1,000 women by age and conception for whites for the year 1972–73 are given in Table 6.4.

Table 6.4   Age-specific-level abortions for whites

Year	Age 12–14	15–17	18–19
1972	1.3	12.7	26.3
1973	1.5	15.4	31.4
1974	1.5	17.4	35.6
1975	1.7	20.1	38.8
1976	1.7	22.2	41.9
1977	1.8	25.1	46.1
1978	1.9	26.9	50.3

Source: Ezzard, Nancy V., et al., Race-specific patterns of abortion use by American Teenagers. *Am. J. Public*, 1982, 72, 809–14.

Table 6.5 Age-specific-level abortions for blacks and other races.

	Age		
Year	12–14	15–17	18–19
1972	4.7	19.0	36.3
1973	6.5	28.1	52.2
1974	7.9	37.7	70.0
1975	9.6	47.8	83.4
1976	10.4	54.8	95.3
1977	10.4	58.3	99.1
1978	10.0	58.5	100.1

(a) Draw the growth curves for the three age groups
(b) Draw the average growth curve (average for all ages)
(c) Assuming linear relationships for legal abortions over time, find the regression coefficients for each age group, and test the hypothesis that these coefficients are the same for the three age groups

2. Table 6.5 gives the same data as in Table 6.4 for black and other races.

(a) Fit a regression line to the average rate
(b) Compare the regression lines for average rates for whites and blacks at .10 level of significance

Statistical Analysis of Growth Curves

Growth curves are used for growth studies in human and animal populations. The growth of children is usually measured by their height and weight at various stages of development. Statistical analysis requires that we have a model of the growth curve. One of the most commonly used models in growth curves is a nonlinear model. It assumes that the variable measuring growth, y, is a linear function of x and log x where x is the age. That is,

$$y = \beta_0 + \beta_1 x + \beta_2 \ln x + e \tag{6.1}$$

where e denotes the random error in the model. Another popular model is:

$$y = \alpha_0 + \alpha_1 x - \exp(r_0 + r_1 x) + e \tag{6.2}$$

The model in Equation (6.1) is linear in parameters and provides estimates of the parameters by the ordinary or weighted least-squares method. The model in Equation (6.2), however, the numerical evaluation of parameters must be determined by computer.

When data on growth are given in terms of multivariate observations, the models become complicated. For a detailed discussion, see Rao (1958).

Many growth experiments use simple analysis-of-variance models. An example is given below from a nutrition experiment on hogs from Wishart (1939), who provided statistical analysis for one of the earliest studies in growth curves.

Example 6.2 (Wishart): In a nutrition experiment, three diets—A = low protein, B = medium protein and C = high protein—were fed to hogs. Each pen contained six animals: 3 hogs and 3 gilts from the same litter. At feeding time, animals were segregated in boxes. Each pen also consisted of 3 heavy and 3 light animals, decided at the time of weaning. The data are given in Table 6.6. For the growth rate, g, the data can be analyzed by an analysis-of-

Table 6.6   Data for Analysis

Pen		Treatment	Sex	w_0	g	h
I	Heavy	A	G	48	9–94	0–199
		B	G	48	10–00	0–146
		C	G	48	9–75	0–136
	Light	C	H	48	9–11	0–139
		B	H	39	8–51	0–154
		A	H	38	9–52	0–209
II	Light	B	G	32	9–24	0–147
		C	G	28	8–66	0–181
		A	G	32	9–48	0–194
	Heavy	C	H	37	8–50	0–144
		A	H	35	8–21	0–119
		B	H	38	9–95	0–178
III	Light	C	G	33	7–63	0–176
		A	G	35	9–32	0–176
		B	G	41	9–34	0–182
	Heavy	B	H	46	8–43	0–171
		C	H	42	8–90	0–155
		A	H	41	9–32	0–176
IV	Heavy	C	G	50	10–37	0–207
		A	H	48	10–56	0–126
		B	G	46	9–68	0–213
	Light	A	G	46	10–98	0–193
		B	H	40	8–86	0–157
		C	H	42	9–51	0–130
V	Light	B	G	37	9–67	0–192
		A	G	32	8–82	0–199
		C	G	30	8–57	0–189
	Heavy	B	H	40	9–20	0–192
		C	H	40	8–76	0–177
		A	H	43	10–42	0–200

Note: w_0 = initial weight at week 0 in pounds
g = average growth rate in lbs per week
$h = \frac{1}{2}$ (rate of change of growth rate in lbs per week)

variance model to see if there is any effect of diet, sex, and pen. The model used here is

$$y_{ijk} = \mu + \alpha_i + \beta_j + \gamma_k + \delta_{ik} + e_{ijkl}$$

with

$$\sum \alpha_i = \sum \beta_j = \sum \gamma_k = \sum_j \delta_{jk} = \sum_k \delta_{jk} = 0$$

The errors are normally distributed with mean 0 and variance σ^2.

$$\alpha_i = \text{Pen effect}$$
$$\beta_j = \text{Diet effect}$$
$$\gamma_k = \text{Sex effect}$$
$$\delta_{jk} = \text{Interaction between diet and sex}$$

The analysis of variance of the growth rate is given in Table 6.7

Since $F_{2,20,.95} = 3.49$, we see that diets are not significantly different at the .05 level. Similarly, sex and interaction are not significantly different at the .05 level. The only significant effect is "pen" at the .05 level, since $F_{4,20,.95} = 2.87$, That is, the only difference found is in the different pens.

An analysis of covariance can be performed by using the initial weight w_0 as a covariate and keeping the rest of the model as before.

Example 6.4: Data on cardiac output on 18 women who were examined at monthly intervals during pregnancy are given by Walters. MacGregor, and Hills (1966). Data on 5 women are given in Table 6.8.

The data are plotted in Figure 6.2 for women A, B, and C. The means are plotted in Figure 6.3. In order to compare the various growth curves, techniques from regression analysis must be used.

Using the week of gestation (x) as an independent variable, the regressions of cardiac output (y) on x are obtained for each woman. We find the following polynomial regressions using the midpoint of the interval of week of gestation

Table 6.7 Analysis of Variance of the Growth Rate

Variation due to	D.F.	Sum of squares	Mean squares	F
Pens	4	4.8518	1.2130	2.918
Diet	2	2.2686	1.1343	2.729
Sex	1	0.4344	0.4344	1.045
Interaction	2	0.4761	0.2380	.572
Error	20	8.3144	0.4157	
Total	29	16.3453		

as the value of x.

$$\text{For}\quad A,\ y = 20.48 - 1.78x + .07x^2$$
$$\text{For}\quad B,\ y = 8.62 - 0.34x + .01x^2$$
$$\text{For}\quad C,\ y = 19.4 - 1.99x + .09x^2$$
$$\text{For}\quad D,\ y = 17.04 - 1.77x + .08x^2$$
$$\text{For}\quad E,\ y = 40.45 - 4.6x + .20x^2$$

Table 6.8   Cardiac Output (liters per minute)

Women

Weeks of gestation	A	B	C	D	E	Mean
12–15	7.4	6.2	5.4	5.8	7.8	6.5
16–19	5.5	5.9	4.9	5.0	7.4	5.7
20–23	5.7	5.6	6.0	5.2	6.6	5.8
24–27	6.6	6.5	4.3	7.7	7.7	6.6
28–31	6.7	6.1	7.8	6.0	9.1	7.1
32–35	4.8	5.7	6.5	5.3	8.9	6.2
36–39	5.6	6.1	6.2	4.0	5.5	5.5

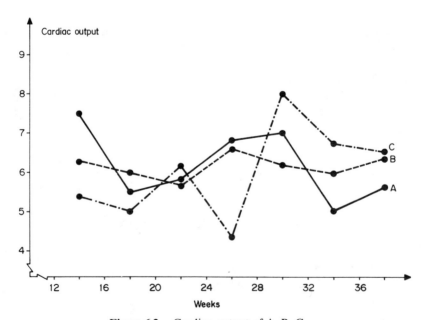

Figure 6.2.   Cardiac output of A, B, C.

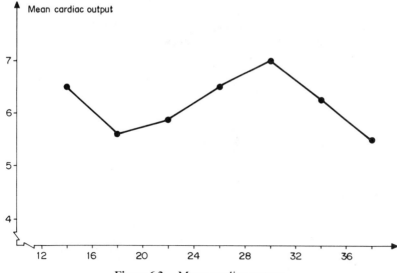

Figure 6.3. Mean cardiac output.

For the means of cardiac output, we have:

$$y = 20.81 - 2.05x + .09x^2$$

Statistical tests of the hypothesis that the growth curves are the same is not rejected at level .05.

Velocity (Rate of Change) of Growth Curves

An interesting question in growth-curve studies is concerned with finding the rate of change of growth. Such concepts lead to *velocity curves*. If the growth curve is given by the equation

$$y = g(t) \tag{6.3}$$

then the *velocity curve* is given by its derivative. Since the data are given at discrete time points, the derivative is approximated by the differences at various time points. That is, if the values of the growth curve at times t_1, t_2 are $g(t_1), g(t_2)$, respectively, then the velocity curve at t_1 is estimated by $v(t_1)$ such that:

$$v(t_1) = \frac{g(t_2) - g(t_1)}{t_2 - t_1} \tag{6.4}$$

Velocity curves provide the measurement for the subject not dependent on initial weight and so forth and are used to compare two or more growth curves. Notice that the velocity estimates are not independent at various time points.

Usually we would like to compare the velocity curve with a horizontal line—
that is, the rate of change has remained constant over the period of study.

Similarly, to study the rate of change of growth curves, the concept of rate of
change of velocity curves is important. This is known as the *acceleration curve*.
If velocity is constant, the acceleration curve will be coincident with the
time axis. At times t_i, the acceleration is given by the rate of change of velocity

Table 6.9   Successive Differences of Means for Example 6.5

Weeks of gestation	Mean cardiac output	Velocity	Acceleration
12–15	6.5		
16–19	5.7	0.8	
20–23	5.8	−0.1	−.9
24–27	6.6	−0.8	−.9
28–31	7.1	−0.5	−1.3
32–35	6.2	.9	1.4
36–39	5.5	.7	−.2

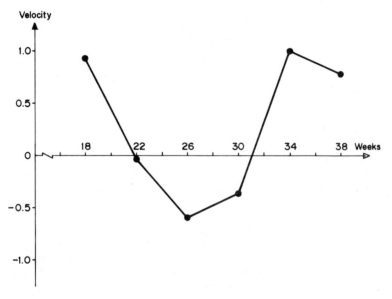

Figure 6.4.   Velocity curve for cardiac output.

between t_{i+1} and t_i:

$$a(t_i) = \frac{v(t_{i+1}) - v(t_i)}{t_{i+1} - t_i} \tag{6.5}$$

Example 6.5: Consider the cardiac output data given in Example 6.4. The velocity curve for the means of the cardiac output is obtained by successive differences of the means. These differences are given in Table 6.9. The data are plotted in Figure 6.4. We take a three-week period as one unit of time so that $t_i - t_{i-1} = 1$ for all i.

Notice that the velocity decreases and then goes up

Statistical Analysis of Incidence Studies

Incidence studies are concerned with studying an event of interest during a certain period. The studies of survival type are incidence studies, since we are studying the event of death. Similarly, studies in reliability where we observe the failure of an item are incidence studies. Survival studies are discussed in Chapter 3.

Exercises

3. Use the analysis-of-covariance model to analyze the data of Example 6.4. Give the model and the hypotheses you are testing.

Table 6.10 Weekly Weight Gain in Grams

	Males	Females
Initial weight	108	100
Week 1	54	35
2	56	24
3	49	18
4	54	21
5	36	13
6	33	11
7	31	12
8	22	8
9	18	5
10	21	11
11	19	4
12	15	4
13	9	5
14	8	−1

Table 6.11   Family Planning Practices in Shanghai County (1971–80)

Year	Eligible couples	Percent using contraceptives	Tubal Ligatia	Vasectomy	IUD	Oral
1971	61,585	68.7	23.409	4,325	7,469	5,073
1972	68,281	72.8	27,251	4,468	8,104	7,156
1973	65,794	82.1	28,441	4,503	9,605	8,646
1974	67,616	85.9	29,465	4,427	10,900	9,906
1975	71,540	84.1	30,151	4,336	11,625	10,384
1976	72,918	85.1	29,629	4,440	13,454	10,619
1977	76,471	85.1	29,604	4,384	15,920	11,249
1978	78,254	86.1	29,998	3,927	19,120	11,324
1979	75,001	87.5	28,115	3,246	24,412	13,048
1980	82,343	98.5	28,125	3,034	30,754	14,953

Source: Xing-Yuan, Gu, and Wei-Sen, Zheng. Family planning, Am. J. Public Health (Supp.). 1982, 72 24–25.

Table 6.12   Average weight of Boys and Girls

Time (days)	Average weight (in ounces)	
	Boys ($n = 14$)	Girls ($n = 13$)
0	100.29	99.62
20	108.29	109.54
40	130.64	131.54
60	150.36	149.77
80	167.14	164.23
100	182.50	176.77
120	198.14	189.85
140	210.93	201.00
160	221.86	209.85
180	233.14	217.69
200	242.43	223.69
220	248.21	228.15
240	255.93	232.69
260	264.57	239.54
280	272.57	246.62
300	280.00	253.46
320	288.07	260.69
340	296.07	269.00

4. The data in Table 6.10 are given for weight gain under two different doses for male and female rats in a 90-day feeding experiment reported by Snee and Acuff (1979).

   (i) Assuming linear growth curves, fit the two growth curves
  (ii) Find if the two growth curves are different
 (iii) Find the velocity curve for males and females

Chapter Exercises

1. Data on family-planning practices in Shanghai County in China during 1971–80 are given in Table 6.11.

   (i) Draw growth curves for the precentage of people using contraceptives
  (ii) Find the linear regression line of this percent on time
 (iii) Fit a second-degree polynomial to the data for the percentage of tubal ligatia and the percentage of vasectomies.
 (iv) Compare the correlation coefficient between the pairs (IUD, oral) and (tubal ligatia, vasectomy).

2. The average weight of boys and girls at intervals of 20 days are given by Rao (1958) in Table 6.12.

   (i) Using an analysis-of-covariance model, test the hypothesis that the growth rate of boys and girls is the same at level .05
  (ii) Obtain the growth curves for boys and girls

Summary

Long-term studies may be *cross-sectional*, where different group of individuals are studied at several instances of time, or *logitudinal*, where the same group is studied over time. Essentially, the longitudinal studies are special cases of time series. When longitudinal studies investigate the occurrence of an event, such as death, during the period, they are called *incidence studies*. We have already considered incidence studies in survival analysis. When we measure the response, such as weight or height of a subject at a given instant of time, we have *growth studies*. *Regression models* are used to analyze growth studies. *Aanalysis of variance* techniques are also applicable to growth experiments. The *velocity curve* of growth provides the rate of change of growth at several instances of time and is used for comparison.

References

Bauman, Karl E.; Anderson, Ann E.; Freeman, Jean L.; and Koch Gary G. Legal abortions and trends in age-specific marriage rates, *Am. J. Public Health*, 1977, *67*, 52–53.

Berkey, Catherine S. Comparison of two longitudinal growth models for preschool children, *Biometrics*, 1982, *38*, 221–34.

Box, G. E. P., and Jenkins, G. M. *Times Series Analysis: Forecasting and Control*, San Francisco: Holden-Day, 1970.

Choi, Keewhan, and Thacker, Stephen D. Mortality during influenza epidemic in the United States, 1967–1978, *Am. J. Public Health*, 1982, *72*, 1280–83.

Goldstein, H. Longitudinal studies and the measurement of change, *The Statistician*, 1968, *18*, 93–117.

Hait, Harvard I.; Lemeshow, Stanley; and Rosenman, Kenneth D. A longitudinal study of blood pressure in a national survey of children, *Am. J. Public Health*, 1982, *72*, 1285–87.

Hemminki, Elma, and Paakkulainen, Anneli, The effect of antibiotic on mortality from infectious diseases in Sweden and Finland, *Am. J. Public Health*, 1976, *66*, 1180–84.

Hills, Michael. *Statistics for Comparative Studies*, New York: John Wiley & Sons, 1974.

Marubini, E.; Resele, L. F., and Barghini, G. A comparative fitting of the Gompertz and logistic functions to longitudinal height data during adolescence in girls, *Human Biology*, 1971, *43*, 237–52.

National Center of Educational Statistics, National Longitudinal Study of the High School Class of 1972. Washington, D.C.: Department of Health, Education and Welfare, 1977.

Pearlman, Alexander W. Breast Cancer: Influence of growth rate on prognosis and treatment evaluation, *Cancer*, 1976, *38*, 1826.

Rao, C. R., Some statistical methods for comparison of growth curves, *Biometrics*, 1958, *14*, 1–17.

Snee, R. D., and Acuff, S. K. A useful method for analysis of growth studies, *Biometrics*, 1979, *35*, 835–48.

Walters, W.; MacGregor, W.; and Hills, M. Cardiac output at rest during pregnancy and puerperium, *Clinical Sciences*, 1966, *30*, 1–11.

Wishart, J. Growth-rate determinations in nutrition studies with the bacon pig and their analysis, *Biometrika*, 1939, *30*, 16–28.

Woodbury, Max A.; Manton, Kenneth G.; and Stallard, Eric, Longitudinal analysis of the dynamics and risk of coronary heart disease in Framingham study, *Biometrics*, 1979, *35*, 575–85.

chapter seven

Statistical Methods in Bioassay

The estimation of the relative potency of an unknown test preparation as compared to a known standard preparation is an important area of biological assay in pharmacology and texicology. Many statistical procedures have been developed to estimate relative potency in bioassay, and the literature on the subject is vast. Some problems in bioassay have resulted in the creation of new areas in statistics, such as stochastic approximation. This chapter is an elementary introduction to some statistical procedures in bioassay. For an extensive treatment of bioassay, see Finney (1964, 1971) and Bliss (1952).

Bioassay-type problems arise in many other fields of application. In engineering applications, such problems are studied under *sensitivity analysis*. For a discussion of sensitivity analysis, see Dixon and Massey (1969).

There are three basic aspects of a bioassay investigation—the stimulus, the subject, and the response. The stimulus such as a drug, vitamin, treatment procedure, or chemical is applied to the subject, which may be an animal, human, insect, or living tissue. The response of the subject is measured for various levels of the stimulus. The purpose of bioassay is to estimate the potency of the stimulus needed to produce the desired response. The figure below shows the stimulus-response relationship. In sensitivity analysis, one is

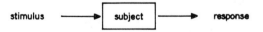

concerned with the estimation of sensitivity to the amount of a given material. For example, to find the amount of explosives in a bomb, one may subject the bomb to various levels of shock to observe whether or not it explodes.

Biological assays are generally performed in two ways. In direct biological assay, the potency of a test preparation can be examined directly, and its relative potency can be determined in comparison with that of a standard preparation. When a substance cannot be assayed directly, we measure the indirect response on the subjects. This indirect response allows us to estimate the relative potency of the test preparation. When the response is measured indirectly, we have *indirect bioassay*; otherwise we have *direct bioassay*. For example, when measuring the relative potency of a vitamin, an indirect measurement must be used, such as the weight gain of a subject under the test and standard preparations of a vitamin, since the potency of a vitamin cannot be measured directly. The response in the indirect assay may be quantal a discrete variable such as yes or no, dead or alive, or it may be quantitative measurement of a variable associated with the subject. We will consider the problem of estimating the potency of a test preparation by direct assay first.

Direct Assays

One of the most common methods in direct assay is the "cat" method. Here the amount of stimulus is continuously given to the subject until a prescribed response is observed. For example, the amount of digitalis is administered just enough to stop the heart of a cat. The ratio of such amounts for standard and test preparations gives an estimate of the relative potency of the drug. It should be remembered that experiments for performing direct assay are not always possible.

Let μ_S be the mean of the standard preparation and μ_T the mean of the test preparation giving the desired response. Then the relative potency of the test preparation is defined by:

$$\rho = \frac{\mu_T}{\mu_S} \tag{7.1}$$

The estimate R of the relative potency can be made in terms of the estimates of μ_S and μ_T. Let $\bar{X}_S$ and $\bar{X}_T$ be the sample averages of standard and test preparations. An estimate of ρ is then given by:

$$R = \frac{\bar{X}_T}{\bar{X}_S} \tag{7.2}$$

The standard deviation of the estimate of R can be obtained for a large sample size. The variance of R is given by (approximately):

$$\text{Var}(R) \simeq \frac{1}{\bar{X}_S^2}[V(\bar{X}_T) + R^2 V(\bar{X}_S)]$$

Recall that the variance of the sample average is the variance of the population divided by the number of observations in the sample. Hence, both $V(\bar{X}_T)$ and $V(\bar{X}_S)$ can be estimated from the sample. If the variance is assumed to be the same for the standard and text preparations, a pooled variance estimate would be used for both.

A confidence-interval estimate of ρ is obtained if we assume that the amount of standard and test preparations needed to get a certain response have normal distributions. If the random variable X has a normal distribution with mean μ and variance σ^2, we denote it by $N(\mu, \sigma^2)$. Suppose the standard preparation has $N(\mu_S, \sigma^2)$ and the test preparation has $N(\mu_T, \sigma^2)$. Let $X_{S1}, X_{S2}, \ldots, X_{Sn}$ be n observations on the standard preparation and $Y_{T1}, Y_{T2}, \ldots, Y_{Tm}$ be m observations on the test preparation. Then the distribution of $\bar{X}_S$ is:

$$N\left(\mu_S, \frac{\sigma^2}{n}\right)$$

and the distribution of $\bar{X}_T$ is:

$$N\left(\mu_T, \frac{\sigma^2}{m}\right)$$

The estimate of the common variance σ^2 is given by:

$$s^2 = (n + m - 2)^{-1}\left[\sum_{i=1}^{n}(X_{Si} - \bar{X}_S)^2 + \sum_{i=1}^{m}(X_{Ti} - \bar{X}_T)^2\right] \tag{7.3}$$

Assuming the independence of the two samples, the distribution of $\bar{X}_T - \rho\bar{X}_S$ is:

$$N\left(0, \sigma^2\left(\frac{1}{n} + \frac{\rho^2}{m}\right)\right) \tag{7.4}$$

since $\mu_T - \rho\mu_S = 0$.

Using s^2 as an estimate of σ^2, then

$$\frac{\bar{X}_T - \rho\bar{X}_S}{s\sqrt{\dfrac{1}{n} + \dfrac{\rho^2}{m}}} \tag{7.5}$$

has a t-distribution with $m + n - 2$ degrees of freedom. The $(1 - \alpha)$-level confidence interval is obtained with the help of the following inequality:

$$-t_{m+n-2, 1-\alpha/2}\, s\sqrt{\frac{1}{n} + \frac{\rho^2}{m}} < \bar{X}_T - \rho\bar{X}_S < t_{m+n-2, 1-\alpha/2}\, s\sqrt{\frac{1}{n} + \frac{\rho^2}{m}} \tag{7.6}$$

$t_{1-\alpha/2}$ is the upper $(1 - \alpha/2)$th percentile of t-distribution with $m + n - 2$ degrees of freedom. Notice that the unknown parameter ρ is involved in both sides of the inequalities. We obtain the interval by solving the quadratic equation:

$$(\bar{X}_T - \rho\bar{X}_S)^2 = t^2 s^2\left(\frac{1}{n} + \frac{\rho^2}{m}\right) \tag{7.7}$$

Equation (7.7) can be rewritten, after multiplying out all terms and collecting those involving ρ^2 and ρ, as:

$$\rho^2\left(\bar{X}_S^2 - \frac{t^2 s^2}{m}\right) - 2\rho\bar{X}_S\bar{X}_T + \bar{X}_T^2 - \frac{t^2 s^2}{n} = 0.$$

Solving it for ρ, we have the two values:

$$\rho = \frac{\bar{X}_S\bar{X}_T \pm \sqrt{\bar{X}_S^2\bar{X}_T^2 - \left(\bar{X}_S^2 - \frac{t^2 s^2}{m}\right)\left(\bar{X}_T^2 - \frac{t^2 s^2}{n}\right)}}{\left(\bar{X}_S^2 - t^2\frac{s^2}{m}\right)} \tag{7.8}$$

If the roots of quadratic equation are not real—that is, the expression in Equation (7.8) inside the radical sign is negative—we are unable to find the confidence interval for the relative potency by this method.

Dividing the numerator and denominator of the right-hand side of (7.8) by $\bar{X}_S^2$, the confidence limits for ρ can also be written as

$$\frac{1}{1-g}\left\{\frac{\bar{X}_T}{\bar{X}_S} \pm \sqrt{\frac{\bar{X}_T^2}{\bar{X}_S^2} - (1-g)\left(1 - \frac{t^2 s^2}{n\bar{X}_T^2}\right)}\right\} \tag{7.9}$$

where

$$g = \frac{t^2 s^2}{m\bar{X}_S^2}$$

The confidence limits given in Equation (7.9) are special cases of the well-known Fieller's theorem in bioassay.

Example 7.1: Let $m = n = 5$, $\bar{X}_S = 2.10$, $\bar{X}_T = 3.15$ and let $s^2 = 0.16$. For a 95% confidence interval, we obtain $t = 2.3$ for 8 degrees of freedom. The estimate of potency is $R = 1.5$:

$$g = \frac{(2.3)^2 \cdot (0.16)}{5(2.1)^2} = 0.038$$

The confidence interval from Equation (19.9) is:

$$(1 - 0.038)^{-1}[1.50 \pm \sqrt{(1.50)^2 - (0.962)(0.983)}]$$

or

$$(0.37, 2.83)$$

When the variances are not assumed to be equal or there is dependence between the samples obtained from the test and the standard preparations, the confidence intervals become more complicated. Expressions for potency in such cases can be found in Finney (1971).

Exercises

1. The standard preparation of a drug gave the determinations of 10.5, 10.7,

9.8, 9.5, and 9.5. For the test preparation, we had 12.7, 11, 5, 10.7, 13.5, 12.7, 13.3:

(a) Find the point estimate of the relative potency of the test preparation.
(b) Give a 90% confidence interval for the relative potency.

2. In a direct assay of a chemical, the test and standard substances gave the determinations:

Test: 1.7,   1.8,   1.9,   2.2,   1.6,   1.9,   1.7
Standard: 1.2,   2.5,   2.3,   1.9,   3.1,   2.1

(a) What is the estimate of the relative potency of the test preparation?
(b) Find a 95% confidence interval for the relative potency.

3. Drugs A, B, and C are being compared for their potencies. Several doses of each drug are given. Find the potency of A as compared to B and the potency of A as compared to C. Find the 95% confidence intervals for their relative potencies:

Drugs	A	B	C
	1.6	2.0	6.2
	1.7	2.3	6.8
	1.4	2.5	5.9
	1.2	1.9	7.8
	1.9	1.8	8.0
	1.5	2.7	6.3
	1.8	2.8	5.7
	2.3	2.3	5.8
	2.5	2.9	
	1.9	1.1	

4. Tinctures A and B are being compared. Find the potency of A relative to B and the 90% confidence interval for it. The doses are:

A	3.4	4.1	2.9	3.7	2.8
B	5.6	7.2	6.9	6.1	8.2

Indirect Assay

In situations where direct assays are not possible, different doses are given to various subjects and the responses are measured. We consider first the case when the response is a continuous variable. The assumption made generally is that the dose-response relationship has a parametric model. We are interested not only in estimating potency but also the overall dose-response relationship and the parameters of the model involved.

In general, very few dose levels can be tested. Various standard designs can be used in conducting a bioassay, since the problem of designing a bioassay is not very different from that of designing any other experiment. The general theory of the design of experiments as discussed in Chapter 11, Volume I applies here as well. We assume that the relationship between the log dose and the response is linear. Let Y_S and Y_T be responses to a dose level of the standard preparation Z_S and Z_t, where $\log Z_S = X_S$ and $\log Z_T = X_T$. We assume linear relations for test and standard preparations

$$Y_T = \alpha_T + \beta_T X_T \tag{7.10}$$

and

$$Y_S = \alpha_S + \beta_S X_S \tag{7.11}$$

If the test preparation has potency ρ, a dose of Z_T (which is the same as ρZ_S) gives the same response Y_S as the standard preparation. Notice that

$$X_T = \log Z_T = \log(\rho Z_S) = \log \rho + \log Z_S = \log \rho + X_S$$

Parallel-Line Assays

Assume first that the dose-response relationship for the test drug is a line with the same slope as the one of the standard preparation. Such assays are called *parallel-line assays*. We then have $\beta_S = \beta_T$ and $Y_T = Y_S$ when $X_T = \log \rho + X_S$, so that

$$Y_S = \alpha_T + \beta_S(\log \rho + X_S)$$

or

$$Y_S = \alpha_T + \beta_S \log \rho + \beta_S X_S \tag{7.12}$$

Comparing Equations (7.11) and (7.12), we have

$$\alpha_S = \alpha_T + \beta_S \log \rho. \quad \text{or} \quad \log \rho = \frac{\alpha_S - \alpha_T}{\beta_S} \tag{7.13}$$

Let $\hat{\alpha}_S, \hat{\alpha}_T$, and $\hat{\beta}_S$ be the estimates of the corresponding parameters. Then:

$$\hat{\alpha}_S = \bar{Y}_S - \hat{\beta}_S \bar{X}_S \tag{7.14}$$

$$\hat{\alpha}_T = \bar{Y}_T - \hat{\beta}_S \bar{X}_T \tag{7.15}$$

$$\hat{\beta}_S = \frac{\sum(X_{Si} - \bar{X}_S)(Y_{Si} - \bar{Y}_S) + \sum(X_{Ti} - \bar{X}_T)(Y_{Ti} - \bar{Y}_T)}{\sum(X_{Si} - \bar{X}_S)^2 + \sum(X_{Ti} - \bar{X}_T)^2} \tag{7.16}$$

The potency ρ can then be estimated in terms $\hat{\alpha}_S, \hat{\alpha}_T$, and $\hat{\beta}_S$ with:

$$\log R = \frac{\hat{\alpha}_S - \hat{\alpha}_T}{\beta_S} \tag{7.17}$$

An alternate form for R is given by:

$$\log R = \frac{\bar{Y}_S - \bar{Y}_T}{\hat{\beta}_S} - (\bar{X}_S - \bar{X}_T) \tag{7.18}$$

We illustrate the procedure of estimation of potency in a parallel-line assay with the help of an example adapted from Finney (1964).

Example 7.2: Suppose an assay of Oestrone is to be made. One standard method is to obtain the weight of the uterus as the percentage of body weight in ovariectomized female rats as a response to standard and test doses after a fixed number of days after treatment. Suppose four litters of 7 animals are used—three animals being used for the standard preparation of 0.2 mg, 0.4 mg, and 0.8 mg, and four animals being used for the test preparation, which consists of doses 0.0075 cc, 0.015 cc, 0.03 cc, and 0.06 cc. The responses, obtained as mg per 100 g of body weight, are given in Table 7.1.

The doses can be standardized so as to obtain simplification in graphing and computations. For example, the standard dose Z_S can be transformed to X_S by dividing by .4 and taking the log to base 2

$$X_S = 2[\log_2 Z_S - \log_2 0.4] \tag{7.19}$$

giving the coded values of X_S as -2, 0, and 2. Similarly,

$$X_T = 2[\log_2 Z_T - \tfrac{1}{2}(\log_2 0.015 + \log_2 0.03)] \tag{7.20}$$

transforms the test preparation to -3, -1, 1, 3.

We obtain the regression equations for the standard and test preparations in coded units using Equations (7.14), (7.15), and (7.16) as follows:

$$Y_S = 73 + 10X_S,$$
$$Y_T = 90 + 10X_T.$$

Using the transformation (7.19) and (7.20), we have:

$$Y_S = 73 - 20\log_2 0.4 + 20\log_2 Z_S$$

and

$$Y_T = 90 - 10(\log_2 0.015 + \log_2 0.03) + 20\log Z_T \tag{7.21}$$

Table 7.1   Daily Dose

Litter	Standard Oestrone			Test Oestrone			
	0.2 μg	0.4 μg	0.8 μg	0.0075 cc	0.015 cc	0.03cc	0.06cc
I	50	80	80	60	90	120	130
II	40	80	100	70	40	60	80
III	60	100	110	50	90	100	120
IV	50	60	70	60	100	120	150
Averages	50	80	90	60	80	100	120

Using Equation (7.17), the estimate of the relative potency is obtained from:

$$\log_2 R = \tfrac{1}{20}[73 - 20\log_2 0.4 - 90 + 10\log_2 0.015 + 10\log_2 0.03]$$
$$= \tfrac{1}{2}[\log_2(0.015 \times 0.03)] - \log_2 0.4 - \tfrac{17}{20}$$

or

$$\log_2 R = \log_2 \frac{\sqrt{0.00045}}{0.4} - 0.85 = \log_2 .05303 - \log_2 1.8025$$

so that

$$R = .029$$

That is, one cc of the test preparation consists of $.03\,\mu g$ of Oestrone. Figure 7.1 gives the regression lines with coded values.

Slope-Ratio Assays

When the dose-response relationship does not lead to parallel lines, we must use other models. We can assume then that the lines have the same intercepts on the y-axis but have different slopes as in Figure 7.2. An assay that uses the ratio of slopes is called the *slope-ratio assays*. In such a case, we can estimate the relative potency of the test preparation with the help of the ratio of the slopes of the lines. For the standard preparation, we have:

$$Y_S = \alpha_S + \beta_S X_S \tag{7.22}$$

For the test preparation, we have:

$$Y_S = \alpha_S + \beta_T(\rho X_S) = \alpha_S + \rho\beta_T X_S \tag{7.23}$$

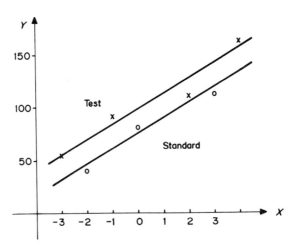

Figure 7.1.   Parallel-line assay.

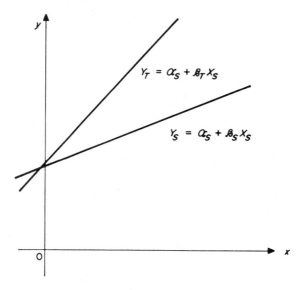

Figure 7.2. Slope-ratio assay.

ρ is the potency of the test preparation. Equations (7.22) and (7.23) give:

$$\rho\beta_T = \beta_S$$

or

$$\rho = \frac{\beta_S}{\beta_T}$$

The estimates of the potency can be made in terms of the ratio of the estimated slopes of the lines:

$$R = \frac{\hat{\beta}_S}{\hat{\beta}_T}$$

Let N_1 be the number of observations from the standard preparation and N_2 be the number of observations from the test preparation. The estimates of β_S, β_T, and α_S, by pooling both samples, are:

$$\hat{\beta}_S = \frac{\sum(X_{Si} - \bar{X}_S)(Y_{Si} - \bar{Y}_S)}{\sum(X_{Si} - \bar{X}_S)^2} \tag{7.24}$$

$$\hat{\beta}_T = \frac{\sum(X_{Ti} - \bar{X}_T)(Y_{Ti} - \bar{Y}_T)}{\sum(X_{Ti} - \bar{X}_T)^2} \tag{7.25}$$

and

$$\hat{\alpha}_S = \frac{N_1(\bar{Y}_S - \hat{\beta}_S \bar{X}_S) + N_2(\bar{Y}_T - \hat{\beta}_T \bar{X}_T)}{N_1 + N_2} \tag{7.26}$$

The formulae for confidence intervals for the potency in parallel-line and slope-ratio assays are not given here and can be found in Finney (1971).

Exercises

5. An extract of adrenal tissue of unknown pressor amine content is assayed by injecting it into a cat. The transient rise in blood pressure is recorded. The following data are obtained from a 2-point assay with 4 observations at each dose. Find the relative potency using parallel-line assay.

Doses

Standard		Test	
$5\,\mu g$	$20\,\mu g$	$0.1\,mg$	$0.4\,mg$
16	37	20	51
15	40	19	49
18	44	23	55
11	39	18	45

6. Suppose the model of the slope-ratio assay is assumed for the data given below. Obtain the relative potency of the test preparation.

Doses

Standard Preparation		Test Preparation	
$5\,\mu g$	$15\,\mu g$	$0.1\,mg$	$0.3\,mg$
17	32	21	46
23	28	20	53
20	30	19	51

7. An assay of vitamin D_3 in cod-liver oil is made through its antiarchitic activity in chickens. The following data (in terms of modified percents) are given as a measure of bone-ash percentages. The standard preparation doses are in B.S.I. units per 100 grams of food and the test preparation doses are in mg C.L.O. per 100 grams of food Finney (1971).

Standard			Test			
5.76	9.6	16	32.4	54	90	150
35	62	116	20	26	57	140
30	67	105	39	60	89	133
24	95	91	16	48	103	142
37	62	94	27	− 8	129	118
28	54	130	− 12	46	139	137
73	56	79	2	77	128	84
31	48	120	31		89	101
21	70	124			86	
− 5	94					
	42					

Use parallel-line assay to determine the potency of the test preparation as compared to the standard preparation.

Probit Analysis

Many situations in bioassay result in quantal responses. In studying the potency of insecticides, the response to various doses of the insecticide are obtained in terms of death or no death of the insects. These responses can be represented by a variable having only two values, 0 and 1. The data at any given dose level are summarized in this case by the proportion of subjects having a specific response. Therefore, in quantal bioassay, the response distribution is to be estimated with the help of observed frequencies of response at various dose levels.

Suppose $F(x)$ is the probability that a dose of a stimulus less than or equal to x produces a response. Let the response distribution be lognormal* with mean μ and variance σ^2. Then in terms of the logarithm of the dose, we can represent $F(x)$ as a probability in terms of the standard normal distribution. The graph of $F(x)$ is the S-shaped curve. To obtain a simple graphical characterization of this curve in the form of a straight line, let y be the upper

---

* A random variable x has a lognormal distribution if log x has the normal distribution. When the mean and variance of log x are μ and σ^2, the mean and variance of x are $\text{Exp}\left(\mu + \dfrac{\sigma^2}{2}\right)$ and $[\text{Exp}(2\mu + 2\sigma^2) - \text{Exp}(2\mu + \sigma^2)]$, respectively.

percentile of the standard normal distribution. Then:

$$y = \frac{x - \mu}{\sigma} = \frac{1}{\sigma}x - \frac{\mu}{\sigma} \qquad (7.27)$$

Equation (7.27) gives a linear relation involving percentiles of the standard normal distribution and the mean and variance of the response distribution. For a given value of $F(x)$, the value of y can be determined from the normal tables. In order to avoid negative values of y for a give value of F, the *probit transformation* is used, in which the value of y is increased by 5. In general the *probit* of a number P is a number $y + 5$, such that P is the probability that the standard normal variable is less than y. Table 7.2 gives a few selected values of the probit transformation.

If the response is death, then the response distribution provides the mortality experience as a function of dose. Some of the parameters, such as quantiles of the response distribution, have become standard quantities for comparison in bioassay.

Median lethal dose. By median lethal dose or its abbreviation, LD_{50}, is meant the median of the response distribution. LD_{50} signifies the value of the dose below which half of the population will die.

Median effective dose. By median effective dose or its abbreviation, ED_{50}, is meant the median of the response distribution when the response is something other than death.

The potency of a test preparation can be compared with that of a standard preparation in terms of its LD_{50}. In some cases, other quantiles of the response distribution, such as LD_{90}, LD_{10}, may be more meaningful for comparison. Figure 7.3 gives a graphical representation of LD_{50} and LD_{90}. Below are given the maximum likelihood estimates for LD_{50} from the data using probits.

Table 7.2   Probit Transformation

Response rate P	0.00	0.01	0.02	0.03	0.04	0.05	0.06	0.07	0.08	0.09
0.00	—	2.67	2.95	3.12	3.25	3.36	3.45	3.52	3.59	3.66
0.10	3.72	3.77	3.82	3.87	3.92	3.96	4.01	4.05	4.08	4.12
0.20	4.16	4.19	4.23	4.26	4.29	4.33	4.36	4.39	4.42	4.45
0.30	4.48	4.50	4.53	4.56	4.59	4.61	4.64	4.67	4.69	4.72
0.40	4.75	4.77	4.80	4.82	4.85	4.87	4.90	4.92	4.95	4.97
0.50	5.00	5.03	5.05	5.08	5.10	5.13	5.15	5.18	5.20	5.23
0.60	5.25	5.28	5.31	5.33	5.36	5.39	5.41	5.44	5.47	5.50
0.70	5.52	5.55	5.58	5.61	5.64	5.67	5.71	5.74	5.77	5.81
0.80	5.84	5.88	5.92	5.95	5.99	6.04	6.08	6.13	6.18	6.23
0.90	6.28	6.34	6.41	6.48	6.55	6.64	6.75	6.88	7.05	7.33

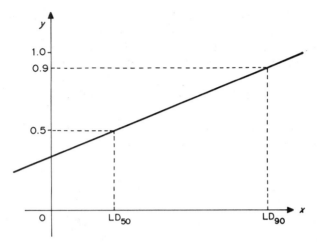

Figure 7.3.   Dose-response relationship.

Let $n_1, n_2, \ldots, n_k$ be the number of subjects at dose levels $z_1, z_2, \ldots, z_k$ respectively. Assume as before that $\log z_i = x_i$, $i = 1, 2, \ldots, k$. Let r_i be the number of subjects giving a response at x_i with probability $p_i, i = 1, 2, \ldots, k$. Suppose the response distribution is normal with mean μ and variance σ^2. Since the probability distribution of r_1 is binomial with parameters n_i and p_i, the likelihood of the sample is given by:

$$L = \prod_{i=1}^{k} \binom{n_i}{r_i} p_i^{r_i} (1 - p_i)^{n_i - r_i} \tag{7.28}$$

Notice that p_i, the probability of a response, depends on the parameters μ and σ^2. The likelihood of the sample given by Equation (7.28) is to be maximized with respect to μ and σ^2. It is not possible to obtain the estimates of μ—that is, of LD_{50}—in closed form. Instead, we must resort to numerical methods. The maximization of Equation (7.28) is to be made under constraints that the p_i's are probits. Computer programs are available for the numerical evaluation of the maximum likelihood estimate of LD_{50}.

Graphical procedures are also available using tabulated values of the weights and working probits. The difficulty of estimating the linear relation in Equation (7.27) is apparent when we notice that the ordinary least-squares method would not suffice. We are fitting a straight line to (y_i, x_i), where y_i is the probit of p_i. The variance of y_i is $p_i(1 - p_i)/n$. Since the variances at each point x_i are different, we need to obtain the least-squares estimates with weights inversely proportional to their variances. But these weights are unknown and hence the difficulty this presents at various stages. For numerical methods for probit analysis, see Finney (1971).

Example 7.3: Suppose doses for the standard preparation (on log scale) are 1, 2, 3, and 4 and for the test preparation are 0, 1, 2, and 3. Let the quantal response be given in terms of the number of positive responses among mice tested on the preparations. The following data are given:

	Standard Preparation			Test Preparation	
Dose	Number of mice	Number responded	Dose	Number of mice	Number responded
1	10	1	0	20	2
2	10	3	1	20	3
3	10	5	2	20	7
4	10	8	3	20	12

For use in probit analysis, we need the probits. The probits for the percentage responses are given in Table 7.2. We have the following results:

Standard Preparation

Dose (x)	Proportion	Probit (y)
1	.1	3.72
2	.3	4.48
3	.5	5.00
4	.8	5.84

Test Preparation

Dose (x)	Proportion	Probit (y)
0	.10	3.72
1	.15	3.96
2	.35	4.61
3	.60	5.25

The regression lines are drawn to the probit in Figure 7.4. Their intercepts and the slope provide the estimate of potency.

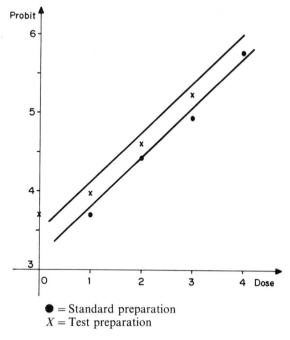

● = Standard preparation
X = Test preparation

Figure 7.4. Probit analysis.

Exercises

8. On log scale, the doses of a standard and test preparation with the number of subjects and corresponding responses are given below. Use probits to fit the linear dose-response relationships and find the estimate of potency of the test preparation from graphs:

	Standard			Test		
Doses	1	2	3	0	1	2
Number of subjects	15	15	15	20	20	20
Number of responses	3	7	11	2	7	10

9. Suppose mice are injected with a dose of insulin and the number of mice with convulsions or symptoms of collapse are recorded. Suppose, on log scale, the doses are measured and we have the following data. Give the dose-

response relationship using probits and find the relative potency of the test preparation from graphs.

	Standard Preparation			Test Preparation		
Doses	.001 (IU)	.002 (IU)	.004 (IU)	20 (IU/mg)	40 (IU/mg)	80 (IU/mg)
Number of mice	10	10	10	10	10	10
Number convulsed	1	2	4	2	3	5

Other Methods for Estimating LD$_{50}$

There are many other methods in addition to probit analysis for estimating potency. Berkson (1944) proposed that a logistic distribution be the response distribution. Then the underlying mathematics can be simplified. Furthermore, the analysis of logistic models compares very well with that of lognormal models. The logistic distribution function is given by:

$$F(x) = \frac{1}{1 + e^{-(\alpha + \beta x)}}, \quad -\infty < x < \infty$$

The logistic distribution results in *logits* compared to *probits* in the normal case. The logit of P is given by $Y = \ln[P/(1 - P)]$. That is:

$$P = \frac{e^Y}{1 + e^Y}$$

The logit of P can be used to find LD$_{50}$ in the same way as the probit has been used. The only difference in using logits is that the response distribution is logistic rather than normal, which results in the use of probits. The logistic distribution for the standard random variable (mean zero and variance one) is very close to the standard normal distribution. A few values of logits are given below.

P	.1	.2	.3	.4	.5	.6	.7
logit P	− 2.197	− 1.386	− .847	− .040	.00	.447	.895

	.8	.9	.98				
	1.450	2.314	4.595				

For a more extensive table of logits, see Berkson (1953).

Two nonparametric methods for estimating LD_{50} are given below.

Spearman-Kärber Method

Suppose $p_1, p_2, \ldots, p_k$ are the proportions of observed responses in a quantal assay at dose levels $x_1, x_2, \ldots, x_k$, respectively. An estimate of the median of the response distribution is obtained by Equation (7.29) if we assume that $p_i, i = 1, 2, \ldots, k$ are in an increasing order.

$$m = \sum_{i=1}^{k-1} (p_{i+1} - p_i)\left(\frac{x_i + x_{i+1}}{2}\right) \tag{7.29}$$

m is known as the *Spearman-Kärber estimate* of LD_{50}. When doses are given at equal intervals—for example, when a log transformation is used with the same dilution ratio—one can use

$$x_{i+1} - x_i = d, \quad i = 1, 2, \ldots, k$$

where d is a known constant. Then m can be simplified by:

$$m_1 = x_k + \tfrac{1}{2}d - d \sum_{i=1}^{k} p_i \tag{7.30}$$

When the number of experimental subjects at each dose x_i is the same, say n, then Equation (7.29) further simplifies to:

$$m_3 = x_k + \tfrac{1}{2}d - \frac{d}{n} \sum_{i=1}^{k} r_i \tag{7.31}$$

The variances of the estimates m, m_1, or m_2 can be obtained using results of binomial distribution. If there are n_i subjects at dose x_i, we know that

$$\mathrm{Var}\left(\frac{r_i}{n_i}\right) = P_i(1 - P_i)/n_i \tag{7.32}$$

so that

$$V(m_1) = d^2 \sum_{i=1}^{k} P_i(1 - P_i)/n_i$$

where P_i is the true probability of the response at x_i. An unbiased estimate of $V(m_1)$ is:

$$\hat{V}(m_1) = d^2 \sum_{i=1}^{k} p_i(1 - p_i)/(n_i - 1) \tag{7.33}$$

Example 7.4: Litchfield and Wilcoxon (1949) give the following data in the assay of an antihistamine drug that was used at various doses to protect test animals against a certainly lethal dose of histamine. Eight animals were used at each level:

Dose $\mu g/kg$	Log Dose	Number of Deaths	p_i
1000	3.0	0	0.00
500	2.7	1	0.12
250	2.4	4	0.50
125	2.1	4	0.50
62.5	1.8	7	0.87

The Spearman-Kärber estimate is:

$$m_1 = 3.0 + \tfrac{1}{2}(0.3) - \frac{0.3}{8}(0 + 1 + 4 + 4 + 7)$$

or

$$m_1 = 2.55$$

The standard deviation of the estimate is:

$$\sqrt{\left\{ \frac{(0.3)^2}{7} [0 + \tfrac{1}{8} \times \tfrac{7}{8} + \tfrac{4}{8} + \tfrac{4}{8} \times \tfrac{4}{8} + \tfrac{7}{8} \times \tfrac{1}{8}] \right\}} = 0.01$$

Up-and-Down Method

A method of obtaining quantal data by performing experiments sequentially, has been of considerable help in sensitivity experiments. Dixon and Mood (1948) introduced the *up-and-down method*, which is also known as the *staircase method* in connection with testing the sensitivity of explosives. For a given dose, one subject is tested and based on the response; the next dose level is chosen. The method requires that a higher level be chosen if there is no response (failure) and a lower level if there is a response (success). Let success be denoted by $\times$ and failure by o. Suppose the levels are given in terms of log dose as $\ldots, h_{-2}, h_{-1}, h_0, h_1, h_2, \ldots$. The results of an experiment are seen like the one in Figure 7.5.

The up-and-down method of experimentation is really designed to provide a good estimate only for LD_{50}. If one wants to estimate other quantiles of the response distribution, especially for the tails of the distribution, such as for LD_{90} and LD_{10}, then this method fails. It is assumed that the dose levels differ

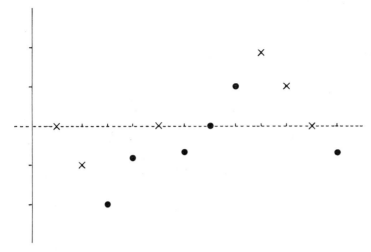

Figure 7.5. Up-and-down method.

by a constant quantity, which is chosen roughly equal to the standard deviation of the response distribution. Since the experimenter has some idea of the response distribution, this assumption is not that stringent.

The number of successes at each level are given:

Dose level	h_{-2}	h_{-1}	h_0	h_1	h_2
Frequency of successes	n_1	n_2	n_3	n_4	n_5

Let $\bar{y}$ be the sample average of the above frequency distribution and s_y^2 the sample variance. Then the estimates of LD_{50} and of standard deviations of the response distribution are given by:

$$LD_{50} = \bar{y} \pm \tfrac{1}{2}d$$

$$s = 1.620d\left(\frac{s_y^2}{d^2} + 0.029\right)$$

The plus sign is used when the analysis is based on 0; otherwise a minus sign is used in LD_{50}.

The estimates in the case of a small sample size and tables to perform such computations are provided by Dixon and Massay (1969).

Example 7.5: Thirteen explosives are dropped from various heights to see if they explode. Explosion (X) and no explosion (0) are shown in Figure 7.5. The frequency distribution of Xs and 0s is given:

Log height	Frequency (X)	Frequency (O)
2	1	1
1.9	1	1
1.8	3	1
1.7	1	3
1.6	—	1

Basing the analysis on Os, we have:

$$\bar{y} = (2 + 1.9 + 1.8 + 5.1 + 1.6)/7 = 12.4/7 = 1.77, \quad d = .1$$

so that $LD_{50} = 1.77 + .05 = 1.82$.

Exercises

10. Solve Exercise 8 using logits.
11. Solve Exercise 9 using logits.
12. Calculate LD_{50} of the standard and test preparations of a drug using the Spearman-Kärber method for the following data. The animals were used at each level:

Number of Responses

Dose (ug/kg)	Standard drug	Dose (cc)	Test drug
5	1	10	2
20	3	30	5
80	7	90	8

13. In an experiment to obtain the LD_{50} of an explosive by up-and-down method, the following data were obtained. Find the LD_{50} and its standard deviation:

Log Dose	Frequency Xs	Os
3	4	—
2.4	10	5
1.8	15	10
1.2	2	12
.6	—	2

Chapter Exercises

1. In a direct assay of a chemical preparation, the following observations were made:

Standard preparation: 11.6,   10.9,   11.8,   10.6,   10.7

Test preparation: 15.7,   15.6,   14.7,   13.6,   15.2

Find the estimate of the relative potency of the test preparation and the 90% confidence interval for it.

2. The following data are obtained for a drug in indirect assay. Find the potency of the estimate using the parallel-line assay technique.

Doses

Standard			Test		
$4\,\mu g$	$12\,\mu g$	$36\,\mu g$	$0.2\,mg$	$0.8\,mg$	$3.2\,mg$
15	18	26	10	25	30
17	14	28	20	15	32
13	20	21	21	20	28
11	15	22	16	24	23

3. The following table was obtained for the effectiveness of an insecticide by the up-and-down method. Find the LD_{50} and its standard deviation. Using normal approximation, find the 95% confidence interval for LD_{50}:

y														
3.1										X	X	X	X
3.0								X	O	O	O	O	
2.9	X		X	X	X			O	O				
2.8	X	O	O	O	O	X	O						
2.7	O				O	O							

Summary

Bioassay or *biological assay* techniques using biological material are used for estimating the *potency* of an unknown test preparation as compared to that of a known preparation. The three elements in bioassay are *subject*, *stimulus*, and *response*. The bioassay may be done by a *direct method*, where the amount of material is measured to provide a given response. *Indirect methods* are used when we are unable to measure the amount to obtain a given response. The indirect assay leads to a *dose-response* relationship. When the response as a function of log dose gives parallel straight lines, it is called *parallel-line assay*. If the lines are not parallel, we have *slope-ratio assay*. When we have a *quantal* response, such as a yes or no, we use *probit method* if the response distribution is assumed to be lognormal. When the response distribution is logistic, the *logit method* is used. There are many nonparametric methods for estimating the median and other percentiles of the response distribution. The median of the response distribution is called ED_{50}, *median effective dose*, or LD_{50}, *median lethal dose*, if the response is lethal. LD_{50} is used to compare response distributions. The *Spearman-Kärber* method and the *up-and-down method* of Dixon and Mood are used when only LD_{50} is to be estimated and lognormality is not assumed.

References

Armitage, P. *Statistical Methods in Medical Research*, New York: John Wiley & Sons, 1971.

Berkson, J. A statistically precise and relatively simple method of estimating the bioassay with quantal response, based on the logistic function, *J. Am. Statist. Assoc.*, 1953, 48, 565–99.

———. Application of the logistic function to biossay. *J. Am. Statist. Assoc.*, 1944, 39, 357–65.

Bliss, C. I. *Statistics in Biology*, New York: McGraw-Hill, 1967.

———. *The Statistics of Bioassay*, New York: Academic Press, 1952.

———, and M. L. Pabst, Assays for standardizing adrenal cortex extract in production, *Bul. Int. Statist. Inst.*, 1955, 34, 317–18.

Dixon, W. J. The up-and-down method for small sample, *J. Am. Statist. Assoc.*, 1965, 60, 967.

———, and Massey, Frank J. *Introduction to Statistical Analysis*, New York: McGraw-Hill, 1969.

———, and Mood, A. M. A method for obtaining and analyzing sensitivity data, *J. Am. Statist. Assoc.*, 1948, 43, 109.

Emmens, C. W. *Principles of Biological Assay*, London: Chapman and Hall, 1948.

———, ed. *Harmone Assay*, New York: Academic Press, 1950.

Finney, D. J. *Probit Analysis*, London: Cambridge University Press, 1971.

———. *Statistical Method in Biological Assay*, 2d ed. New York: Hafner, 1971.

chapter eight

Statistical Analysis of Compartmental Systems

In pharmacology, physiology, biochemistry, and several other areas of the biomedical sciences, many problems are concerned with the study of rates of exchange of material between the components of a system. In biological contexts, such components are called *compartments*. Compartmental models play a central role in pharmacokinetics, which is the study of the kinetics of the absorption, distribution, metabolism, and excretion of drugs.

The use of compartmental models in pharmacokinetics has increased because regulating agencies and medical societies have become more concerned with the therapeutic and toxic responses of drugs in humans. Also, the importance of the availability of a drug for absorption—generally known as *bioavailability* which is the proportion of a drug administered but not excreted—has become an important criterion for differentiating between two drugs.

Many other areas of scientific study, such as economics, engineering, and ecology, employ compartmental models and the literature is considerable. Atkins (1969) and Jacquez (1972) give a comprehensive account of the mathematics of compartmental models with special reference to the biomedical sciences.

A model by its very nature is a simple and crude approximation of the phenomenon. A *stochastic model* describes a phenomenon in terms of random quantities with associated probability distributions. Stochastic models lead to the study of a given phenomenon in terms of systems of random variables. The building of stochastic models will not be pursued here. This chapter will describe *deterministic models* for compartmental systems in order to develop methods for finding estimates for the constants that are contained in the models. These models involve variables that contain no probabilistic elements, and they are generally given in terms of well-known functions or differential equations.

Kinds of Compartmental Models

Suppose we want to determine the amount of an atmospheric pollutant retained in the human body when the rates of uptake and excretion are given. In this case, the body can be regarded as a single compartment and the system can be studied with the help of mathematical equations. A one-compartment model is shown in Figure 8.1.

When we want to determine whether the substance is in blood or tissue, for example, we have two compartments—one for blood and the other for tissue. Suppose the substance is first given in blood, such as by an intravenous injection, and is excreted through urine. Then we may have a two-

Figure 8.1. One-compartment model.

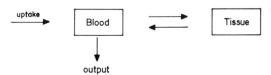

Figure 8.2. Two-compartment model.

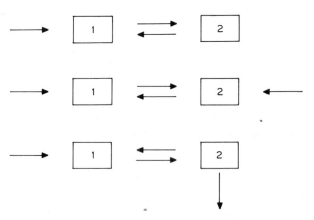

Figure 8.3. Two-compartment models.

compartment system involving blood and tissue compartments (Fig. 8.2). We assume here that the exchange between blood and tissue compartments and output is only through the blood compartment.

There are many possibilities in two-compartment systems. There may be input into both compartments, or there may be output from the second compartment. Three situations are shown in Figure 8.3.

A compartmental system is said to be closed when no material can enter it or leave it. For example, a system where, once injected, the drug metabolizes within the system without being excreted is a *closed system*. Otherwise, it is an *open system*.

The system may be *catenary* or *mammilary*. In a catenary system involving *p* compartments, the compartments have nonzero exchange rates only with the compartments adjacent to them. A catenary system of three compartments is shown in Figure 8.4.

In a *mammilary system* with *p* compartments, *p* − 1 compartments have exchange rates only with a central compartment and not among themselves. This system for 4 compartments is shown in Figure 8.5.

Example 8.1 (catenary system): An I^{131} labeled albumin distribution has been studied by Berson and Yalow (1954) with the help of a three-compartment catenary system having a plasma compartment, a rapidly exchanging extravascular compartment, and a slowly changing extravascular compartment with input into the plasma and output from the extravascular compartment.

Example 8.2 (mammilary system): Consider plasma in the body as the central compartment. Consider material that is injected into plasma and various organs in the body as the other compartments so that the interchange

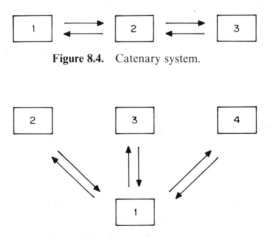

Figure 8.4.   Catenary system.

Figure 8.5.   Mammilary system.

of the material is only between the plasma and on organ. In this way the plasma and other organs form a mammilary system.

Exercises

1. Give an example in which the third two-compartment system in Figure 8.3 has input in compartment 1, output from compartment 2, and an exchange between them.
2. To evaluate the 'body burden' of a substance to which a human being is exposed—for example, particulate matter in the atmosphere—a compartmental model can be used. How would you describe the model? How many compartments does it have?
3. Suppose a substance is metabolized in two organs of the body and we want to develop a model for the 'organ burden' of an atmospheric contaminant. How many compartments are needed? Make a diagram giving input and output, assuming that the output is through feces and the intake is through the lungs.

Mathematical Models of Compartmental Systems

When the rates of exchange between compartments or between compartments and the environment are constants, the dynamics of the system involving these rates usually can be expressed with the help of linear differential equations. Since difference equations are discrete analogs of differential equations, we can use difference equations to study the mathematical models of the compartmental systems.

Suppose we consider the one-compartment system where the rate of output is constant k. That is, the material in the compartment at time t, denoted by y_t, is excreted at the rate of k per unit of amount y_t in the system. We assume that there is no continuous input into the system. Such a one-compartment model is shown in the schematic diagram:

The amount in the compartment at time $t + 1$ is y_{t+1}. It is decreased by the amount ky_t. Therefore,

$$y_{t+1} = y_t - ky_t \tag{8.1}$$

gives the relationship of k and y_t. Equation (8.1) connects successive amounts in the compartment with the help of the equation:

$$y_{t+1} = (1 - k)y_t \tag{8.2}$$

The ratio of the successive amounts at times $t + 1$ and t is the constant $1 - k$. If we write Equation (8.1) as

$$y_{t+1} - y_t = -ky_t$$

then it shows that the rate of change of the amount at time t is $-ky_t$. Equation (8.2) is generally known as a difference equation.

Suppose the initial amount in the compartment is y_0. Then successive amounts at times $1, 2, \ldots$ can be obtained by using Equation (8.2). Notice that

$$y_1 = (1 - k)y_0$$
$$y_2 = (1 - k)y_1 = (1 - k)(1 - k)y_0 = (1 - k)^2 y_0$$
$$y_3 = (1 - k)y_2 = (1 - k)^3 y_0$$

and so on. The solution of the difference equation (8.2) can be written as

$$y_t = (1 - k)^t y_0 \qquad (8.3)$$

When k and y_0 are known, the amount at any time t can be determined.

When the time is taken as a continuous variable, the rate of change can be expressed as a derivative. An approximation of Equation (8.3) is given by:

$$y_t = y_0 \exp(-kt) \qquad (8.4)$$

Equation (8.4) is used often in applications.

We do not know the initial amount and the rate k in general. Experiments performed on a compartment result in observations $y_1, y_2, \ldots$ at various times $t = 1, 2, \ldots$. Statistical analysis of the data is needed to obtain estimates of y_0 and k. The next section will consider the estimation problem. Notice that the problem is that of regression analysis and can be solved by well-known methods see Chapter 8, Volume I.

Example 8.3: The excretion rate of penicillin is assumed to be governed by the mathematical relation (8.4). The initial dose is y_0. We find that at time $t = \dfrac{1}{k}$, the concentration of the drug will be $\dfrac{1}{e}$ times the initial dose. That is, $1 - \dfrac{1}{e}$ of the drug, approximately 63.2 percent (based on an approximate value of $e = 2.718$) of the drug has disappeared by time $\dfrac{1}{k}$. Suppose a constant amount of drug y_0 is added at intervals of time T. Then at time T we have the total amount:

$$y_0 e^{-kT} + y_0$$

At time $2T$, we have:

$$y_0 + (y_0 + y_0 e^{-kT})e^{-kT}$$

And so on for any time nT:

$$y_0 + (y_0 + y_0 e^{-kT} + \cdots + y_0 e^{-k(n-1)T})e^{-kT} \tag{8.5}$$

Equation (8.5) is a geometric series and its sum is:

$$\frac{y_0(i - e^{-(n+1)kT})}{1 - e^{-Tk}} \tag{8.6}$$

As n becomes large, the concentration approaches

$$\frac{y_0}{1 - e^{-kT}} \tag{8.7}$$

since the limit of $e^{-(n+1)kT}$ as n approaches infinity is equal to zero. In order to achieve the concentration in Equation (8.7), one gives a booster dose equal to:

$$y_1 = \frac{y_0}{1 - e^{kT}} \tag{8.8}$$

In this case, the amount of drug disappearing is equal to:

$$y_1(1 - e^{-kT}) = y_0$$

The constant infusion of amount y_0 brings it back to y_1. This shows that the initial dose should be $e/(e - 1)$ times the repeated dose, $\dfrac{e}{e - 1}$ is approximately 1.6. This is used in practice when a physician prescribes two tablets initially and one thereafter, since 1.6 is approximated by 2 tablets.

Example 8.4: Suppose a single dose of a drug is given through injection and its concentration in plasma is measured at various times. Then we have a one-compartment (plasma) model given by Equation (8.4).

Half-life of a Drug

Let y_0 be the initial dose of a drug. The time at which half of the drug, $y_0/2$, remains in the system is called the *half-life* of the drug. Half-life is used to compare several drugs or several methods of administration of the drug, such as oral or intravenous. Assume that the system has one compartment. Thus the concentration at any given time is given by:

$$y = y_0 e^{-kt} \tag{8.9}$$

Now $y = \dfrac{y_0}{2}$ at $t = t_{1/2}$, the half-life of the drug. So that

$$\frac{y_0}{2} = e^{-kt_{1/2}} \tag{8.10}$$

giving

$$t_{1/2} = \frac{\ln 2}{k} = \frac{.693}{k}$$

If the half-life is known, the constant k in the model can be obtained by:

$$k = \frac{\ln 2}{t_{1/2}} = \frac{.693}{t_{1/2}}$$

Suppose the initial dose of a drug is 200 mg and the constant $k = .173$ (hours)$^{-1}$. Then the half-life of the drug is:

$$t_{1/2} = \frac{.693}{.173} = 4 \text{ hours}$$

One-Compartment Models with Nonconstant Rates

When the rates of exchange are based on some other assumptions, the models cannot be described in terms of simple linear differential or difference equations. Therefore, these solutions are not available in terms of an exponential function such as Equation (8.4). An important elimination scheme, known as the Michaelis-Menten elimination scheme (Fig. 8.6), arises in enzyme kinetics as well as in pharmacokinetics. Here we have another constant, V_m, giving the maximum reaction rate and k_m, the Michaelis constant. The model is represented by:

Figure 8.6.   Michaelis-Menten scheme.

The model is described by:

$$y_{t+1} - y_t = -\frac{V_m y_t}{k_m + y_t}$$

This scheme is useful in the study of certain drugs (see Wagner, 1973). For this mathematical model, y_t cannot be explicitly expressed in terms of t. When we are given observations $y_1, y_2, \ldots$ at times $t = 1, 2, \ldots$, constants k_m and V_m can be estimated by numerical methods.

Two-Compartment Models

The system consisting of two compartments with exchange rates k_1 and k_2 between them and k_3 as the rate of output from one of the compartments is shown by Figure 8.7.

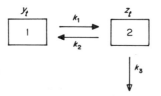

Figure 8.7. Two-compartment model.

At any given time t, let y_t be the concentration in compartment 1 and z_t in compartment 2. Therefore, at time $t + 1$, the concentration in compartment 1 is reduced by the amount $k_1 y_t$ and increased by the amount $k_2 z_t$. That is:

$$y_{t+1} = y_t - k_1 y_t + k_2 z_t \qquad (8.11)$$

Similarly, for compartment 2, we have:

$$z_{t+1} = z_t + k_1 y_t - k_2 z_t - k_3 z_t \qquad (8.12)$$

Compartmental equations provide the concentrations in terms of rates. When observations are made only on one compartment, say compartment 1, then we can eliminate z_t from the above equations and can get an equation involving y_t, y_{t+1}, and y_{t+2}. This leads to a second-order difference equation of the model. The treatment of other kinds of rates of exchange between two compartments can be similarly studied. For reference, see Atkins (1969) and Jacques (1972).

Example 8.5:   Let $k_1 = .15$, $k_2 = .1$, and $k_3 = .2$. Then Equations (8.11) and (8.12) are:

$$y_{t+1} = .85y_t + .1z_t \qquad (8.13)$$

and

$$z_{t+1} = .7z_t + .15y_t$$

Changing t to $t + 1$, we have from Equation (8.13):

$$y_{t+2} = .85y_{t+1} + .1z_{t+1}$$

Substituting the values of z_t and z_{t+1}:

$$y_{t+2} = .85y_{t+1} + .1(.7z_t + .15y_t)$$
$$y_{t+2} = .85y_{t+1} + .7(y_{t+1} - .85y_t) + .015y_t$$

or

$$y_{t+2} = 1.15y_{t+1} - .58y_t$$

So the second-order difference equation representing the model is:

$$y_{t+2} - 1.55y_{t+1} + .58y_t = 0$$

In general, a two-compartment system with constant rates of exchange between compartments leads to the following model, combining two exponential terms:

$$y = Ae^{Bt} + Ce^{Dt}$$

This function is the solution of a second-order linear differential equation, which results when we have observations on only one compartment.

The concentration at any given time t in a general c compartment system connected with constant rates of exchange, input and output, is given by a linear combination of c exponential terms:

$$y = \sum_{i=1}^{c} A_i e^{B_i t} \tag{8.14}$$

The solutions of many important compartmental problems are reduced in terms of Equation (8.14). Therefore, fitting a linear function of exponentials to data is important in compartmental analysis. Numerical procedures for fitting the sum of exponentials to experimental data are available through statistical packages such as SAS and BMDP.

Exercises

4. Suppose a two-compartment model has a constant rate k_1 of input into compartment 1, flow from compartment 1 to compartment 2 at the rate k_{12}, and from compartment 2 to compartment 1 at the rate k_{21}. The output is also from compartment 1 at rate k_3. Give the compartmental system diagram and derive the difference equations for the model.
5. In a one-compartment system, there is an instant input of 150 mg of a drug. Suppose the rate of excretion is such that the half-life of the drug is 4 hours. Find the model equation. What is the amount of drug in the body after 8 hours?
6. In a two-compartment system (Fig. 8.6), let $k_1 = .10$, $k_2 = .2$, and $k_3 = .15$. Find the equations of the model for concentrations in the compartment. If observations were made only on compartment 1, what would the equation of the model be?

Statistical Analysis

In a practical situation, the experimenter does not know the rate constants, so the experiment must determine them. The comparison of rate constants for two different drugs may throw some light on their bioavailability. Also, one may be able to find out the dose levels of the drug to be prescribed so as to attain a certain concentration.

Consider Equation (8.4). Let the observations at times $1, 2, \ldots, n$, be y_1, $y_2, \ldots, y_n$. The assumed model with natural logarithmic transformation is

$$\ln y_t = \ln y_0 - kt + \varepsilon_t \tag{8.15}$$

where ε_t, $t = 1, 2, \ldots, n$ are random errors with means zero and variances σ^2. Let

$$\ln y_t = Y_t$$

and

$$\ln y_0 = a$$

Then we have Equation (8.12) as a simple linear model:

$$Y_t = a - kt + \varepsilon_t$$

The estimates of a and k can be obtained with the help of the least-squares method, with:

$$\hat{a} = \bar{Y} - \hat{k}\bar{t}$$

$$\hat{k} = \frac{\sum(Y_t - \bar{Y})(t - \bar{t})}{\sum(t - \bar{t})^2}$$

The theory of simple linear regression was discussed in Chapter 9, Volume I, and all the results hold in the case of the above compartmental model.

Using the alternative model in terms of difference equations, we have:

$$y_{t+1} = (1 - k)y_t + \eta_t \tag{8.16}$$

Here the errors η_t are again assumed to be random with means 0 and variances σ^2. Notice that the errors in the Equations (8.8) and (8.9) have different interpretations. In obtaining estimates of k from Equation (8.9), we use weighted least-squares estimates. Suppose the weights are w_t. Then the estimate is given by:

$$\hat{k} = 1 - \frac{\sum w_t y_t y_{t+1}}{\sum w_t y_t^2} \tag{8.17}$$

In ordinary situations, we assume that w_t are known, using the values of the reciprocals of the estimated variances at t as the weights. Sometimes when there is only one observation at t, $w_t = \dfrac{1}{y_t}$ gives a fairly good estimate of k.

Note that Equation (8.9) is the model of time series. Such models have been studied extensively in the literature. See, for example, Box and Jenkins (1970).

Example 8.6: Suppose the concentration of a drug in the body is given by a one-compartment system as in Equation (8.4). The concentrations at various times are given:

Time (hours)	0	1	2	4	6	8	10	12	14	16
Plasma concentration $(\mu g/ml)$	99	82	67	44	30	20	14	9	6	4

Fitting the model

$$Y = \ln y = \ln y_0 - kt$$

we then have the following:

t	0	1	2	4	6	8	10	12	14	16
Y	4.59	4.41	4.20	3.78	3.40	3.00	2.64	2.20	1.79	1.39

$$\sum t_i = 73, \quad \sum Y_i = 31.40, \quad \sum t_i^2 = 817$$
$$\sum t_i Y_i = 172.43$$
$$\hat{k} = \frac{172.43 - (73)(31.40)/10}{817 - (73)^2/10} = -.2$$
$$\ln \hat{y}_0 = 3.14 + 1.46 = 4.6$$
$$y_0 = 99.48$$

Hence the fitted model is:

$$y = 99.48e^{-.2t}$$

Graphical Procedure

Using semilogarithmic paper, we can draw a straight line to the scatter plot of the data for fitting the one-compartment model in Equation (8.8). This procedure is quite common in practice. For more than one compartment, the method of *peeling* can be used. Suppose we want to fit a model with three exponential terms. The object is to fit the three terms sequentially. Suppose the straight line drawn on a semilogarithmic paper provides the fitting to one term. Then the residuals are plotted on a semilogarithmic paper and the second term is fitted to the residuals by a straight line. Residuals from the second can be used to fit the third term.

Number of Compartments

We are sometimes not sure whether the pharmacokinetics of a drug should be based on–one- or two-compartment model. Statistically, this can be determined by testing hypotheses about the constants in the model. Or we can fit the data sequentially to one exponential term. If the test of residuals determines a good fit to the model, we can assume that there is only one compartment. If not, we can fit the second term as well. Computer procedures for such algorithms are available.

Exercises

7. For the following data, fit a one-compartment model:

t	0	1	2	4	6	8	10
y	5.1	4.5	3.9	3.5	2.7	1.9	1.7

What is the concentration at $t = 3$?

8 Blood concentrations of methylene blue in dogs following rapid intravenous injection of 2 mg/kg dose are given by DiSanto and Wagner (1972):

t = Time (hours)	.046	.175	.325	.493	.997	2.20	3.11	4.22
y = Concentration $\left(\dfrac{\mu g}{ml}\right)$	.941	.365	.275	.199	.124	,0640	.0424	.0310

Fit a one-compartment model to the data and obtain the residuals. Fit another
one-compartment model to the residuals. Test the hypothesis that the
coefficient of the second term is zero at the .05 level of significance.

Chapter Exercises

1. Counts describing the decay of the neutron density in a medium-size
   assembly of beryllium are given by Cornell (1962), as quoted from de
   Saussure and Silver. Fit the model in Equation (8.4) to the modified data:

t	0	2	4	6	8	10	12	14	16
y_t	100,145	60,305	36,205	21,705	13,045	7,835	4,782	2,915	1,752

2. Plasma concentrations of the drug Isoxicam are given to men after oral and
   rectal administration in various doses in Table 8.1. Fit a model with two
   compartments to the oral-administration group using weighted least-
   squares estimates.
3. Smolen and Schoenfeld (1956) studied the phenomenon of the drug
   Tropicamide absorption in rabbits. Intravenous injections of the drug
   (50 mg per kilogram of weight) were given to four rabbits and pupillary
   diameters were measured. A function of Mydriatic Response Intensity (I)
   was calculated:

Time in (minutes)	0	10	20	30	40	50	60	70
$y_t = f(I)$	48.1,	24.7,	12.7,	6.2,	3.1,	1.4,	0.7,	0.3

Fit a two compartment model for this phenomena.

Table 8.1 Plasma concentration (μg^*ml) of Isoxicam after oral or rectal administration n = 11

Time after does (hr)	200 mg (personal)		200 mg (rectal)		400 mg (rectal)	
1	.90	.57	.47	.14	.56	.27
2	1.85	.71	1.01	.39	1.39	.67
4	3.91	1.14	2.39	.78	3.61	1.59
6	5.18	1.26	3.34	.78	5.31	1.75
8	6.10	1.51	4.03	.89	6.87	1.24
10	6.69	1.74	4.78	1.73	8.13	1.97
12	6.72	1.51	5.18	1.80	9.05	2.40
24	5.70	2.0	4.45	2.40	8.42	3.74
48	3.51	.83	2.50	1.34	4.96	2.69
76	2.10	.52	1.51	.81	2.75	1.44
96	1.19	.37	0.94	.51	1.55	.94
120	0.82	.27	0.62	.37	.98	.62

Source: Accardo, S., *et al.* Pharmacokinetics and relative bioavailability of isoxicam after oral and rectal administration in man, *Current Therapeutic Research*, 1983, *33*, 976–81.

Summary

Compartment models are used in many applications. In pharmacokinetics, they are used to obtain the *bioavailability* of drugs. In this chapter, deterministic models of one- and two-compartments were discussed. When rates of exchange between compartments are constants, they lead to the model of concentration in terms of a sum of exponentials. These models can be fitted with the help of the least-squares method. Tests of hypotheses about the number of compartments are made through techniques of regression analyses.

References

Atkins, G. L. *Multicompartment Models for Biological Systems*, London: Methuen, 1969.
Berson, S. A., and Yalow, R. S. The distribution of I^{131}-labeled human serum albumin introduced into ascitic fluid: Analysis of the kinetics of a three-compartment catenary transfer system in man and speculation on possible sites of degradation, *J. Clin. Invest.*, 1954, *33*, 377–87.
Box, George E. P., and Jenkins, Gwilym M. *Time Series Analysis: Forecasting and Control*, San Francisco: Holden-Day, 1970.
Cornell, R. G. A method of fitting linear combinations of exponentials, *Biometrics*, 1962, *18*, 104–13.
DiSanto, A. R. and Wagner, J. G. Pharmacokinetics of highly ionized drugs III. Methylene blue blood levels in dogs and tissue levels in the rat following intravenous administration, *J. Pharm. Sci.*, 1972, *61*, 1090–94.
Godfrey, Keith. *Compartment Models and Their Applications*, New York: Academic Press, 1984.
Jacquez, John A. *Compartmental Analysis in Biology and Medicine*, New York: Elsevier, 1972.
Rodda, B. E.; Simpson, C. B.; and Smith, D. W. The one-compartment open model: Some statistical aspects of parameter estimation, *Appl. Statist.*, 1975, *24*, 309–18.

Rubinow, S. I., and Winzer, A. Compartment analysis: An inverse problem. *Math. Biosc.*, 1971, *11*, 203–47.

Rustagi, J. S. Mathematical models in medicine, *International Journal of Mathematical Education in Science and Technology*, 1971, *2*, 193–203.

———. Mathematical models of body burden. *Arch. Environ. Health*, 1976, *10*, 761–67.

———, and Singh, Umed, *Statistical Analysis of Compartmental Models with Applications to Pharmacokinetics and Bioavailability, Proceedings of The Applied Statistics Conference*, P. R. Krishnaiah, ed. New York: Academic Press, 1977, pp. 461–80.

Smolen, V. F., and Schoenweld, R. D. Drug-absorption analysis from pharmacological data I: Method and confirmation exemplified for mydriatic drug Tropicamide, *J. Pharm. Sci.*, 1971, *60*, 96–103.

Storstein, L. Studies on digitalis XI: Digitoxin metabolism in patients with impaired renal function, *Clinical Pharmacology and Therapeutics*, 1977, *21*, 536–46.

———, and Janssen, H. Studies on digitalis VI: The effect of heparin on serum binding of digitoxin and digoxon. *Clinical Pharmacology and Therapeutics*, 1976, *20*, 15–23.

Wagner, J. G. Properties of the Michaelis-Menten equation and its integrated from which are useful in pharmacokinetics, *Journal of Pharmacokinetics and Biopharmaceutics*, 1973, *1*, 103–21.

———. *Fundamentals of Clinical Pharmacokinetics*, Hamilton, Ill.: Drug Intelligence Publications, 1979.

chapter nine

Sequential Techniques

Statistical procedures usually are based on a fixed sample size. In many situations, the decision to observe an additional observation can be made to depend on the result of the previous observations. Such experiments, which are conducted in a sequence, form a part of statistics known as *sequential analysis*. Sequential analysis is concerned with the design and analysis of an experiment in which a stopping rule is incorporated. In scientific research, many investigators follow such a strategy unconsciously. Scientists experiment until they find definite evidence to support their hypothesis.

The experiment may give an apparent form of a sequential experiment, but in reality it may not be so. In a clinical trial, patients may enter the study sequentially, but there may be no rule of *stopping* the trial based on the results of previous trials. In the application of sequential analysis, one has to make a decision whether to stop or to continue experimentation at each stage.

By performing sequential experiments, the investigator saves on the number of observations. When experimentation is expensive or when the observations are made over long periods, such a saving is necessary. With a given fixed sample size, controlling both types of errors in testing hypotheses is not possible. However, they can be controlled by sequential experimentation. There are many practical situations where sequential experimentation does not work. In the follow-up studies of chronic diseases or in evaluating the effects of treatments that give a response after a long period, the results of the experiment that are needed to make decisions about taking an additional observation are not available.

The application of sequential methods was discussed in Chapters 4 and 5 in the context of medical trials and sampling inspection. This chapter discusses the general method of testing hypotheses when the observations are taken in a sequence. The optimal property of a sequential test is that, for the same probabilities of errors, the sequential test proposed saves observations as compared to a fixed sample size text. The operating characteristic function of the test as well as the mathematical expressions for average sample number function are given. Several applications are also discussed.

Sequential Probability Ratio Test

To test a simple hypothesis against a simple alternative, suppose the probability density is $f(x, \theta)$ and we test the null hypothesis:

$$H: \quad \theta = \theta_0$$
$$A: \quad \theta = \theta_1$$

We assume that the probability of rejecting H when it is true is given as α and the probability of accepting H when H is not true is given as β. α, β are known.

When the sample size is fixed in advance, it is not possible to obtain a test that controls both probabilities α and β. For this reason Neyman and Pearson formulated the problem as a one of finding a test of size α, which maximizes the power. Several tests were given for the above hypothesis for a fixed sample size. The sequential test proposed by Wald (1947) is similar to the likelihood ratio test and is called the *sequential probability ratio test* (SPRT).

Suppose we have m observations $x_1, x_2, \ldots, x_m$ at stage m. Let:

$$\lambda_m = \frac{f(x_1, \theta_1)f(x_2, \theta_1) \ldots f(x_m, \theta_1)}{f(x_1, \theta_0)f(x_2, \theta_0) \ldots f(x_m, \theta_0)} \tag{9.1}$$

The SPRT test is given in terms of λ_m by the following rule involving constants A and B.

Decision Rule:

(i) Continue sampling if

$$B < \lambda_m < A \tag{9.2}$$

where A and B are specified in terms of α and β.

(ii) If $\lambda_m \geq A$, stop sampling and reject H.
(iii) If $\lambda_m \leq B$, stop sampling and accept H.

A and B are approximately determined by:

$$A = \frac{1 - \beta}{\alpha} \tag{9.3}$$

$$B = \frac{\beta}{1 - \alpha} \tag{9.4}$$

The rule is now completely specified.

An interesting property of the sequential probability ratio test is that one should eventually be able to reject or accept the hypothesis H. Wald showed that when the number of observations is large, the sampling terminates with probability one.

Sequential test of a proportion. Suppose in a population, the incidence of a disease is p. Using sequential sampling, we want to test the hypothesis with given α and β:

$$H: \quad p = p_0$$
$$A: \quad p = p_1$$

Let X_i be one if the person tested has the disease and zero otherwise. The probability density of X_i is given by:

$$p^x(1-p)^{1-x}, \quad x = 0, 1 \tag{9.5}$$

When we have taken m observations:

$$\lambda_m = \frac{p_1^{\Sigma X_i}(1-p_1)^{m-\Sigma X_i}}{p_0^{\Sigma X_i}(1-p_0)^{m-\Sigma X_i}} \tag{9.6}$$

Let $\sum_{i=1}^{m} X_i = y$ and consider the logarithm of λ_m:

$$\log \lambda_m = y \log\left(\frac{p_1}{p_0}\right) + (m-y)\log\frac{1-p_1}{1-p_0}$$

The rule is given in the following:

(i) Continue sampling if:

$$\log B < y\log\frac{p_1}{p_0} + (m-y)\log\frac{1-p_1}{1-p_0} < \log A$$

(ii) Reject H if:

$$y\log\frac{p_1}{p_0} + (m-y)\log\frac{1-p_1}{1-p_0} \geq \log A$$

(iii) Accept H if:

$$y\log\frac{p_1}{p_0} + (m-y)\log\frac{1-p_1}{1-p_0} \leq \log B$$

Notice that in terms of y and m the above inequalities divide the region in (m, y)-plane in three parts determined by the following straight lines:

$$y\left(\log\frac{p_1}{p_0} - \log\frac{1-p_1}{1-p_0}\right) + m\log\frac{1-p_1}{1-p_0} = \log A \tag{9.7}$$

and

$$y\left(\log\frac{p_1}{p_0} - \log\frac{1-p_1}{1-p_0}\right) + m\log\frac{1-p_1}{1-p_0} = \log B \tag{9.8}$$

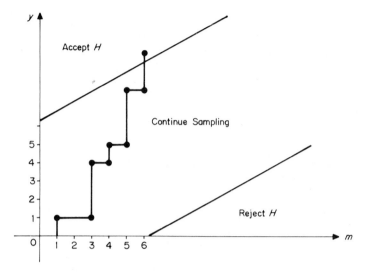

Figure 9.1. Sampling path of a sequential scheme.

Notice that Equations (9.7) and (9.8) are parallel lines and the rule can be described geometrically showing the rejection, acceptance, and continuation regions.

Figure 9.1 shows the sampling path of the above sequential sampling scheme. At each stage of experimentation, using y and m, we can determine the decision to be taken according to the region the point (y, m) falls in. As soon as the path crosses either of the boundaries, the sampling is terminated and corresponding decision is taken. Notice that y and m are discrete variables. The continuous boundaries are shown for simplicity. The geometrical representation is simple and the rule is easy to carry out. In many production lines such as in pharmaceutical industry, sequential analysis is frequently used for quality-control purposes. Such plans of unrestricted sampling are called *open plans*.

To avoid the possibility of too many observations if the sampling continues over an extended period, stopping rule specifying the maximum number of observations to be taken are frequently used. Such plans are called *truncated*, *restricted*, or *closed sampling plans*. The truncation number may be determined with the help of error probabilities. In medical trials, see Armitage (1960) for closed sequential plans in medical trials.

Example 9.1:   With $\alpha = \beta = .05$, we test the hypothesis for the binomial parameter p of the following nature.

$$H: \quad p = 0.4$$
$$A: \quad p = 0.8$$

Substituting $p_0 = 0.4$, $p_1 = 0.8$ and $A = \dfrac{1 - .05}{.05} = 19$, $B = \dfrac{.05}{1 - .05} = \dfrac{1}{19}$, in Equations (9.7) and (9.8), the boundary of the test regions are given by:

$$y(\log 2 - \log \tfrac{1}{3}) + m \log \tfrac{1}{3} = \log 19$$

and

$$y(\log 2 - \log \tfrac{1}{3}) + m \log \tfrac{1}{3} = \log \tfrac{1}{19}$$

That is:

$$0.778y - 0.477m = 1.279$$
$$0.778y - 0.477m = -1.279$$

The boundaries for the test regions then can be graphed in (y, m)-plane.

Exercises

1. Test sequentially the hypothesis for a proportion p, where $\alpha = \beta = 0.02$. Graph the boundaries of the regions.

$$H: \quad p = \tfrac{1}{3}$$
$$A: \quad p = \tfrac{1}{2}$$

2. Suppose $f(x, \theta) = \theta e^{-x\theta}$, $\theta > 0$. Find the sequential probability ratio test at $\alpha = \beta = 0.04$ levels for the hypothesis:

$$H: \quad \theta = 2$$
$$A: \quad \theta = 4$$

Give the graphical regions.

3. For the binomial distribution, test sequentially the following hypotheses.

$$H: \quad p = .1$$
$$A: \quad p = .3$$

Given $\alpha = .10$ and $\beta = .05$, give the SPRT and graph the boundaries of the regions.

4. In Exercise 2, test the following hypothesis using a sequential probability ratio test for $\alpha = .10$ and $\beta = .20$:

$$H: \quad \theta = 1$$
$$A: \quad \theta = 2$$

Graph the rejection and acceptance regions.

Operating Characteristic Function (OC) and Average Sample Number Function (ASN)

The performance of a sequential test can be measured by the operating characteristic function (OC). The probability of accepting the null hypothesis when it is not true is called the *operating characteristic function*. It was shown in Chapter 7, Volume I that the power function of a test is related to the operating characteristic:

Power Function $= 1 -$ Operating Characteristic Function

When two tests are to be compared, the OC function provides the comparison. In a sampling inspection, the operating characteristic provides the probability of accepting the lot when the proportion-defectives in the lot is a given number p.

Example 9.2: In a sequential sampling where we want to estimate the proportion-defectives in a lot, we test the following hypothesis at the producer and consumer risks, α and β, respectively:

$$H: \quad p = p_0$$
$$A: \quad p = p_1$$

The operating characteristic, $L(p)$, of the sequential probability ratio test given by Equations (9.7) and (9.8) has the following values at $p = 0, p_0, p_1$, and 1.

p	$L(p)$
0	1
p_0	$1 - \alpha$
p_1	β
1	0

One more point between p_0 and p_1 can be obtained by:
At

$$p = \frac{\log\dfrac{1 - p_1}{1 - p_0}}{\log\dfrac{1 - p_1}{1 - p_0} - \log\dfrac{p_1}{p_0}}, \quad L(p) = \frac{\log\dfrac{1 - \beta}{\alpha}}{\log\dfrac{1 - \beta}{\alpha} - \log\dfrac{\beta}{1 - \alpha}}$$

Suppose we have $p_0 = .1$, $p_1 = .2$, and $\alpha = \beta = .05$. Then we have:

p	0	.1	.15	.2	1
$L(p)$	1	.95	.50	.05	0

The graph of this OC function is shown in Figure 9.2.

The OC function provides the probability of accepting the lot when the value of proportion-defectives in the lot is p. The operating characteristic functions are also used for comparison of different sampling inspection procedures.

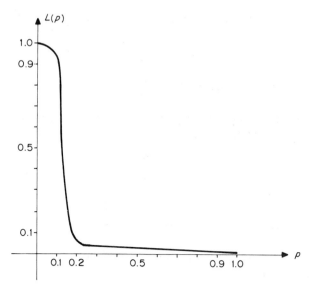

Figure 9.2.   Operating characteristic function.

Average Sample Number

The number of observations required to perform a given test of a hypothesis with the preassigned probabilities of error is a random variable. The expected value of this random variable is called the *average sample number function*. The mathematical form for the average sample function is given below for the case of normal distribution.

If the null and alternative hypotheses for the mean θ are

$$H: \quad \theta = \theta_0$$
$$A: \quad \theta = \theta_1$$

we have the average sample number under the null hypothesis as:

$$E_0(n) = \frac{(1 - \alpha) \ln B + \alpha \ln A}{-\frac{1}{2}(\theta_0 - \theta_1)^2}$$

Under the alternative hypothesis, we have:

$$E_1(n) = \frac{\beta \ln B + (1 - \beta) \ln A}{\frac{1}{2}(\theta_0 - \theta_1)^2}$$

Here A and B are given by Equation (9.3) and (9.4).

Exercises

5. Give five points on the OC function for testing the hypothesis about the

number of proportion-defectives in a lot with $\alpha = .05$ and $\beta = 10$:

$$H: \quad p = .05$$
$$A: \quad p = .10$$

Graph the OC function and from the graph provide the probability of accepting the lot if $p = .15$.

6. Find the average sample number under the null hypothesis when we test the mean of a normal distribution θ, having $\alpha = .05$ and $\beta = .07$, with:

$$H: \quad \theta = .1$$
$$A: \quad \theta = .3$$

Sequential Test for the Mean of the Normal Distribution

Assume that the variance, σ^2, of the normal distribution is known and we want to test the hypothesis sequentially:

$$\text{versus} \qquad \begin{aligned} H: &\quad \mu = \mu_0 \\ A: &\quad \mu = \mu_1 \end{aligned}$$

The sequential probability ratio test requires the computation of λ_m:

$$\lambda_m = \frac{\left(\dfrac{1}{\sqrt{2\pi}\sigma}\right)^m e^{-\frac{1}{2\sigma^2}\Sigma(X_i - \mu_0)^2}}{\left(\dfrac{1}{\sqrt{2\pi}\sigma}\right)^m e^{-\frac{1}{2\sigma^2}\Sigma(X_i - \mu_1)^2}}$$

$$\ln \lambda_m = \frac{1}{2\sigma^2}\left[\sum(X_i - \mu_1)^2 - \sum(X_i - \mu_0)^2\right]$$

$$= \frac{1}{2\sigma^2}\left[\sum X_i^2 - 2\mu_1\sum X_i + m\mu_1^2 - \sum X_i^2 + 2\mu_0\sum X_i - m\mu_0^2\right]$$

$$\ln \lambda_m = \frac{1}{2\sigma^2}\left[2(\mu_0 - \mu_1)\sum X_i + m(\mu_1^2 - \mu_0^2)\right]$$

Or $\ln \lambda_m = a_0 y + b_0$, where:

$$y = \sum_{i=1}^{m} X_i, \quad a_0 = \frac{\mu_0 - \mu_1}{\sigma^2}, \quad b_0 = \frac{m(\mu_1^2 - \mu_0^2)}{2\sigma^2}$$

Hence, we have the decision rule:

    (i) Continue sampling if $\ln B < a_0 y + b_0 < \ln A$
    (ii) If $a_0 y + b_0 \geq \ln A$, stop sampling and reject H.
    (iii) If $a_0 y + b_0 \leq \ln B$, stop sampling and accept H.

Table 9.1 Average Percent Saving in Sequential Sampling for the Mean of a Normal

β	0.01		0.02		0.03		0.04		0.05	
0.01	58	58	54	60	51	61	49	62	47	63
0.02	60	54	56	56	53	57	50	58	49	59
0.03	61	51	57	53	54	54	51	55	50	55
0.04	62	49	58	50	55	51	52	52	50	53
0.05	63	47	59	49	55	50	53	50	51	51
True hypothesis	H	A	H	A	H	A	H	A	H	A

Source: Wald (1947).

Constant A and B are determined by Equations (9.3) and (9.4) in terms of α and β.

Notice that the rejection and continuation regions can be described in terms of straight lines drawn on the axes of y and m, as in the binomial case shown in Figure 9.1.

Saving of Observations

The expected number of observations required by sequential sampling using the above decision rule is much smaller than the most powerful test based on a fixed sample size. Table 9.1 shows the average percentage of savings in a sample. For example, if $\alpha = .01$ and $\beta = .04$ and the null hypothesis is true, the saving is 62 percent by using the sequential probability ratio test. Notice that for the common error levels in Table 9.1, there is at least a 47 per cent saving.

Chapter Exercises

1. Graph the rejection, continuation, and acceptance regions for testing the following hypothesis about the mean of the normal distribution with variance 25, with $\alpha = .03$ and $\beta = .05$:

$$\text{versus} \quad \begin{array}{l} H: \quad \mu = 3 \\ A: \quad \mu = 12 \end{array}$$

What is the percentage of saving of observations under the two hypotheses?
2. What are the boundaries of the sequential test regions for testing the hypothesis about the proportion of diseased individuals in a population, for the hypotheses: $H: p = .03$, $A: p = .04$, with $\alpha = \beta = .10$? Graph the boundaries.
3. What is the OC function of the procedure in Exercise 2?
4. It is assumed that the weights of cereal boxes filled by a machine are

normally distributed with variance $4(\text{grams})^2$. On a production line, a test is being made sequentially for the mean weight at $\alpha = \beta = .05$:

$$H: \quad \mu = 50$$
$$A: \quad \mu = 53$$

What are the boundaries of the regions of acceptance and rejection of the hypothesis H? What is the average sample number when the null hypothesis is true? What is the percentage of saving of observations with the sequential test?

Summary

Sequential analysis is concerned with the sampling scheme where obervations are taken in a sequence. For testing a simple hypothesis versus the simple alternative, we use Wald's *sequential probability ratio rest* (SPRT). This test provides the decision rule in terms of the *region of acceptance* of the null hypothesis, the *region of rejection* of the null hypothesis, and the *continuation region*. If the observed sample statistic is in the continuation region, additional observations are needed. The *operating characteristic function* of a test provides the probability of accepting the hypothesis when the value of the parameter is given. The *average sample number* is the expected sample size in a sequential experiment. In many cases, there is a considerable saving of observations under sequential sampling.

References

Armitage, P. *Sequential Medical Trials.* Springfield, Ill: Charles C. Thomas, 1960.
Wald, A. *Sequential Analysis.* New York. John Wiley & Sons, 1947.
Wetherill, G. B. *Sequential Methods in Statistics.* London: Methuen, 1966.

Appendix Tables

Table I Individual Terms, Binomial Distribution

n	x	.05	.10	.15	.20	.25	.30	.35	.40	.45	.50
1	0	.9500	.9000	.8500	.8000	.7500	.7000	.6500	.6000	.5500	.5000
	1	.0500	.1000	.1500	.2000	.2500	.3000	.3500	.4000	.4500	.5000
2	0	.9025	.8100	.7225	.6400	.5625	.4900	.4225	.3600	.3025	.2500
	1	.0950	.1800	.2550	.3200	.3750	.4200	.4550	.4800	.4950	.5000
	2	.0025	.0100	.0225	.0400	.0625	.0900	.1225	.1600	.2025	.2500
3	0	.8574	.7290	.6141	.5120	.4219	.3430	.2746	.2160	.1664	.1250
	1	.1354	.2430	.3251	.3840	.4219	.4410	.4436	.4320	.4084	.3750
	2	.0071	.0270	.0574	.0960	.1406	.1890	.2389	.2880	.3341	.3750
	3	.0001	.0010	.0034	.0080	.0156	.0270	.0429	.0640	.0911	.1250
4	0	.8145	.6561	.5220	.4096	.3164	.2401	.1785	.1296	.0915	.0625
	1	.1715	.2916	.3685	.4096	.4219	.4116	.3845	.3456	.2995	.2500
	2	.0135	.0486	.0975	.1536	.2109	.2646	.3105	.3456	.3675	.3750
	3	.0005	.0036	.0115	.0256	.0469	.0756	.1115	.1536	.2005	.2500
	4	.0000	.0001	.0005	.0016	.0039	.0081	.0150	.0256	.0410	.0625
5	0	.7738	.5905	.4437	.3277	.2373	.1681	.1160	.0778	.0503	.0312
	1	.2036	.3280	.3915	.4096	.3955	.3602	.3124	.2592	.2059	.1562
	2	.0214	.0729	.1382	.2048	.2637	.3087	.3364	.3456	.3369	.3125
	3	.0011	.0081	.0244	.0512	.0879	.1323	.1811	.2304	.2757	.3125
	4	.0000	.0004	.0022	.0064	.0146	.0284	.0488	.0768	.1128	.1562
	5	.0000	.0000	.0001	.0003	.0010	.0024	.0053	.0102	.0185	.0312
6	0	.7351	.5314	.3771	.2621	.1780	.1176	.0754	.0467	.0277	.0156
	1	.2321	.3543	.3993	.3932	.3560	.3025	.2437	.1866	.1359	.0938
	2	.0305	.0984	.1762	.2458	.2966	.3241	.3280	.3110	.2780	.2344
	3	.0021	.0146	.0415	.0819	.1318	.1852	.2355	.2765	.3032	.3125
	4	.0001	.0012	.0055	.0154	.0330	.0595	.0951	.1382	.1861	.2344
	5	.0000	.0001	.0004	.0015	.0044	.0102	.0205	.0369	.0609	.0938
	6	.0000	.0000	.0000	.0001	.0002	.0007	.0018	.0041	.0083	.0156

(Continued)

Table I (Continued)

n	x						p				
		.05	.10	.15	.20	.25	.30	.35	.40	.45	.50
7	0	.6983	.4783	.3206	.2097	.1335	.0824	.0490	.0280	.0152	.0078
	1	.2573	.3720	.3960	.3670	.3115	.2471	.1848	.1306	.0872	.0547
	2	.0406	.1240	.2097	.2753	.3115	.3177	.2985	.2613	.2140	.1641
	3	.0036	.0230	.0617	.1147	.1730	.2269	.2679	.2903	.2918	.2734
	4	.0002	.0026	.0109	.0287	.0577	.0972	.1442	.1935	.2388	.2734
	5	.0000	.0002	.0012	.0043	.0115	.0250	.0466	.0774	.1172	.1641
	6	.0000	.0000	.0001	.0004	.0013	.0036	.0084	.0172	.0320	.0547
	7	.0000	.0000	.0000	.0000	.0001	.0002	.0006	.0016	.0037	.0078
8	0	.6634	.4305	.2725	.1678	.1001	.0576	.0319	.0168	.0084	.0039
	1	.2793	.3826	.3847	.3355	.2670	.1977	.1373	.0896	.0548	.0312
	2	.0515	.1488	.2376	.2936	.3115	.2965	.2587	.2090	.1569	.1094
	3	.0054	.0331	.0839	.1468	.2076	.2541	.2786	.2787	.2568	.2188
	4	.0004	.0046	.0185	.0459	.0865	.1361	.1875	.2322	.2627	.2734
	5	.0000	.0004	.0026	.0092	.0231	.0467	.0808	.1239	.1719	.2188
	6	.0000	.0000	.0002	.0011	.0038	.0100	.0217	.0413	.0703	.1094
	7	.0000	.0000	.0000	.0001	.0004	.0012	.0033	.0079	.0164	.0312
	8	.0000	.0000	.0000	.0000	.0000	.0001	.0002	.0007	.0017	.0039
9	0	.6302	.3874	.2316	.1342	.0751	.0404	.0207	.0101	.0046	.0020
	1	.2985	.3874	.3679	.3020	.2253	.1556	.1004	.0605	.0339	.0176
	2	.0629	.1722	.2597	.3020	.3003	.2668	.2162	.1612	.1110	.0703
	3	.0077	.0446	.1069	.1762	.2336	.2668	.2716	.2508	.2119	.1641
	4	.0006	.0074	.0283	.0661	.1168	.1715	.2194	.2508	.2600	.2461
	5	.0000	.0008	.0050	.0165	.0389	.0735	.1181	.1672	.2128	.2461
	6	.0000	.0001	.0006	.0028	.0087	.0210	.0424	.0743	.1160	.1641
	7	.0000	.0000	.0000	.0003	.0012	.0039	.0098	.0212	.0407	.0703
	8	.0000	.0000	.0000	.0000	.0001	.0004	.0013	.0035	.0083	.0176
	9	.0000	.0000	.0000	.0000	.0000	.0000	.0001	.0003	.0008	.0020

n	x										
10	0	.5987	.3487	.1969	.1074	.0563	.0282	.0135	.0060	.0025	.0010
	1	.3151	.3874	.3474	.2684	.1877	.1211	.0725	.0403	.0207	.0098
	2	.0746	.1937	.2759	.3020	.2816	.2335	.1757	.1209	.0763	.0439
	3	.0105	.0574	.1298	.2013	.2503	.2668	.2522	.2150	.1665	.1172
	4	.0010	.0112	.0401	.0881	.1460	.2001	.2377	.2508	.2384	.2051
	5	.0001	.0015	.0085	.0264	.0584	.1029	.1536	.2007	.2340	.2461
	6	.0000	.0001	.0012	.0055	.0162	.0368	.0689	.1115	.1596	.2051
	7	.0000	.0000	.0001	.0008	.0031	.0090	.0212	.0425	.0746	.1172
	8	.0000	.0000	.0000	.0001	.0004	.0014	.0043	.0106	.0229	.0439
	9	.0000	.0000	.0000	.0000	.0000	.0001	.0005	.0016	.0042	.0098
	10	.0000	.0000	.0000	.0000	.0000	.0000	.0000	.0001	.0003	.0010
11	0	.5688	.3138	.1673	.0859	.0422	.0198	.0088	.0036	.0014	.0004
	1	.3293	.3835	.3248	.2362	.1549	.0932	.0518	.0266	.0125	.0055
	2	.0867	.2131	.2866	.2953	.2581	.1998	.1395	.0887	.0513	.0269
	3	.0137	.0710	.1517	.2215	.2581	.2568	.2254	.1774	.1259	.0806
	4	.0014	.0158	.0536	.1107	.1721	.2201	.2428	.2365	.2060	.1611
	5	.0001	.0025	.0132	.0388	.0803	.1321	.1830	.2207	.2360	.2256
	6	.0000	.0003	.0023	.0097	.0268	.0566	.0985	.1471	.1931	.2256
	7	.0000	.0000	.0003	.0017	.0064	.0173	.0379	.0701	.1128	.1611
	8	.0000	.0000	.0000	.0002	.0011	.0037	.0102	.0234	.0462	.0806
	9	.0000	.0000	.0000	.0000	.0001	.0005	.0018	.0052	.0126	.0269
	10	.0000	.0000	.0000	.0000	.0000	.0000	.0002	.0007	.0021	.0054
	11	.0000	.0000	.0000	.0000	.0000	.0000	.0000	.0000	.0002	.0005
12	0	.5404	.2824	.1422	.0687	.0317	.0138	.0057	.0022	.0008	.0002
	1	.3413	.3766	.3012	.2062	.1267	.0712	.0368	.0174	.0075	.0029
	2	.0988	.2301	.2924	.2835	.2323	.1678	.1088	.0639	.0339	.0161
	3	.0173	.0852	.1720	.2362	.2581	.2397	.1954	.1419	.0923	.0537
	4	.0021	.0213	.0683	.1329	.1936	.2311	.2367	.2128	.1700	.1208
	5	.0002	.0038	.0193	.0532	.1032	.1585	.2039	.2270	.2225	.1934
	6	.0000	.0005	.0040	.0155	.0401	.0792	.1281	.1766	.2124	.2256
	7	.0000	.0000	.0006	.0033	.0115	.0291	.0591	.1009	.1489	.1934
	8	.0000	.0000	.0001	.0005	.0024	.0078	.0199	.0420	.0762	.1208
	9	.0000	.0000	.0000	.0001	.0004	.0015	.0048	.0125	.0277	.0537
	10	.0000	.0000	.0000	.0000	.0000	.0002	.0008	.0025	.0068	.0161
	11	.0000	.0000	.0000	.0000	.0000	.0000	.0001	.0003	.0010	.0029
	12	.0000	.0000	.0000	.0000	.0000	.0000	.0000	.0000	.0001	.0002

(Continued)

Table I (Continued)

n	x						p					
		.05	.10	.15	.20	.25	.30	.35	.40	.45	.50
13	0	.5133	.2542	.1209	.0550	.0238	.0097	.0037	.0013	.0004	.0001
	1	.3512	.3672	.2774	.1787	.1029	.0540	.0259	.0113	.0045	.0016
	2	.1109	.2448	.2937	.2680	.2059	.1388	.0836	.0453	.0220	.0095
	3	.0214	.0997	.1900	.2457	.2517	.2181	.1651	.1107	.0660	.0349
	4	.0028	.0277	.0838	.1535	.2097	.2337	.2222	.1845	.1350	.0873
	5	.0003	.0055	.0266	.0691	.1258	.1803	.2154	.2214	.1989	.1571
	6	.0000	.0008	.0063	.0230	.0559	.1030	.1546	.1968	.2169	.2095
	7	.0000	.0001	.0011	.0058	.0186	.0442	.0833	.1312	.1775	.2095
	8	.0000	.0000	.0001	.0011	.0047	.0142	.0336	.0656	.1089	.1571
	9	.0000	.0000	.0000	.0001	.0009	.0034	.0101	.0243	.0495	.0873
	10	.0000	.0000	.0000	.0000	.0001	.0006	.0022	.0065	.0162	.0349
	11	.0000	.0000	.0000	.0000	.0000	.0001	.0003	.0012	.0036	.0095
	12	.0000	.0000	.0000	.0000	.0000	.0000	.0000	.0001	.0005	.0016
	13	.0000	.0000	.0000	.0000	.0000	.0000	.0000	.0000	.0000	.0001
14	0	.4877	.2288	.1028	.0440	.0178	.0068	.0024	.0008	.0002	.0001
	1	.3593	.3559	.2539	.1539	.0832	.0407	.0181	.0073	.0027	.0009
	2	.1229	.2570	.2912	.2501	.1802	.1134	.0634	.0317	.0141	.0056
	3	.0259	.1142	.2056	.2501	.2402	.1943	.1366	.0845	.0462	.0222
	4	.0037	.0349	.0998	.1720	.2202	.2290	.2022	.1549	.1040	.0611
	5	.0004	.0078	.0352	.0860	.1468	.1963	.2178	.2066	.1701	.1222
	6	.0000	.0013	.0093	.0322	.0734	.1262	.1759	.2066	.2088	.1833
	7	.0000	.0002	.0019	.0092	.0280	.0618	.1082	.1574	.1952	.2095
	8	.0000	.0000	.0003	.0020	.0082	.0232	.0510	.0918	.1398	.1833
	9	.0000	.0000	.0000	.0003	.0018	.0066	.0183	.0408	.0762	.1222
	10	.0000	.0000	.0000	.0000	.0003	.0014	.0049	.0136	.0312	.0611
	11	.0000	.0000	.0000	.0000	.0000	.0002	.0010	.0033	.0093	.0222
	12	.0000	.0000	.0000	.0000	.0000	.0000	.0001	.0005	.0019	.0056
	13	.0000	.0000	.0000	.0000	.0000	.0000	.0000	.0001	.0002	.0009
	14	.0000	.0000	.0000	.0000	.0000	.0000	.0000	.0000	.0000	.0001

n	k										
15	0	.4633	.2059	.0874	.0352	.0134	.0047	.0016	.0005	.0001	.0000
	1	.3658	.3432	.2312	.1319	.0668	.0305	.0126	.0047	.0016	.0005
	2	.1348	.2669	.2856	.2309	.1559	.0916	.0476	.0219	.0090	.0032
	3	.0307	.1285	.2184	.2501	.2252	.1700	.1110	.0634	.0318	.0139
	4	.0049	.0428	.1156	.1876	.2252	.2186	.1792	.1268	.0780	.0417
	5	.0006	.0105	.0449	.1032	.1651	.2061	.2123	.1859	.1404	.0916
	6	.0000	.0019	.0132	.0430	.0917	.1472	.1906	.2066	.1914	.1527
	7	.0000	.0003	.0030	.0138	.0393	.0811	.1319	.1771	.2013	.1964
	8	.0000	.0000	.0005	.0035	.0131	.0348	.0710	.1181	.1647	.1964
	9	.0000	.0000	.0001	.0007	.0034	.0116	.0298	.0612	.1048	.1527
	10	.0000	.0000	.0000	.0001	.0007	.0030	.0096	.0245	.0515	.0916
	11	.0000	.0000	.0000	.0000	.0001	.0006	.0024	.0074	.0191	.0417
	12	.0000	.0000	.0000	.0000	.0000	.0001	.0004	.0016	.0052	.0139
	13	.0000	.0000	.0000	.0000	.0000	.0000	.0001	.0003	.0010	.0032
	14	.0000	.0000	.0000	.0000	.0000	.0000	.0000	.0000	.0001	.0005
	15	.0000	.0000	.0000	.0000	.0000	.0000	.0000	.0000	.0000	.0000
20	0	.3585	.1216	.0388	.0115	.0032	.0008	.0002	.0000	.0000	.0000
	1	.3774	.2702	.1368	.0576	.0211	.0068	.0020	.0005	.0001	.0000
	2	.1887	.2852	.2293	.1369	.0669	.0278	.0100	.0031	.0008	.0002
	3	.0596	.1901	.2428	.2054	.1339	.0716	.0323	.0123	.0040	.0011
	4	.0133	.0898	.1821	.2182	.1897	.1304	.0738	.0350	.0139	.0046
	5	.0022	.0319	.1028	.1746	.2023	.1789	.1272	.0746	.0365	.0148
	6	.0003	.0089	.0454	.1091	.1686	.1916	.1712	.1244	.0746	.0370
	7	.0000	.0020	.0160	.0545	.1124	.1643	.1844	.1659	.1221	.0739
	8	.0000	.0004	.0046	.0222	.0609	.1144	.1614	.1797	.1623	.1201
	9	.0000	.0001	.0011	.0074	.0271	.0654	.1158	.1597	.1771	.1602
	10	.0000	.0000	.0002	.0020	.0099	.0308	.0686	.1171	.1593	.1762
	11	.0000	.0000	.0000	.0005	.0030	.0120	.0336	.0710	.1185	.1602
	12	.0000	.0000	.0000	.0001	.0008	.0039	.0136	.0355	.0727	.1201
	13	.0000	.0000	.0000	.0000	.0002	.0010	.0045	.0146	.0366	.0739
	14	.0000	.0000	.0000	.0000	.0000	.0002	.0012	.0049	.0150	.0370
	15	.0000	.0000	.0000	.0000	.0000	.0000	.0003	.0013	.0049	.0148
	16	.0000	.0000	.0000	.0000	.0000	.0000	.0000	.0003	.0013	.0046
	17	.0000	.0000	.0000	.0000	.0000	.0000	.0000	.0000	.0002	.0011
	18	.0000	.0000	.0000	.0000	.0000	.0000	.0000	.0000	.0000	.0002
	19	.0000	.0000	.0000	.0000	.0000	.0000	.0000	.0000	.0000	.0000
	20	.0000	.0000	.0000	.0000	.0000	.0000	.0000	.0000	.0000	.0000

Table II   Individual Terms, Poisson Distribution

m

x	0.1	0.2	0.3	0.4	0.5	0.6	0.7	0.8	0.9	1.0
0	.9048	.8187	.7408	.6703	.6065	.5488	.4966	.4493	.4066	.3679
1	.0905	.1637	.2222	.2681	.3033	.3293	.3476	.3595	.3659	.3679
2	.0045	.0164	.0333	.0536	.0758	.0988	.1217	.1438	.1647	.1839
3	.0002	.0011	.0033	.0072	.0126	.0198	.0284	.0383	.0494	.0613
4	.0000	.0001	.0002	.0007	.0016	.0030	.0050	.0077	.0111	.0153
5	.0000	.0000	.0000	.0001	.0002	.0004	.0007	.0012	.0020	.0031
6	.0000	.0000	.0000	.0000	.0000	.0000	.0001	.0002	.0003	.0005
7	.0000	.0000	.0000	.0000	.0000	.0000	.0000	.0000	.0000	.0001

m

x	1.1	1.2	1.3	1.4	1.5	1.6	1.7	1.8	1.9	2.0
0	.3329	.3012	.2725	.2466	.2231	.2019	.1827	.1653	.1496	.1353
1	.3662	.3614	.3543	.3452	.3347	.3230	.3106	.2975	.2842	.2707
2	.2014	.2169	.2303	.2417	.2510	.2584	.2640	.2678	.2700	.2707
3	.0738	.0867	.0998	.1128	.1255	.1378	.1496	.1607	.1710	.1804
4	.0203	.0260	.0324	.0395	.0471	.0551	.0636	.0723	.0812	.0902
5	.0045	.0062	.0084	.0111	.0141	.0176	.0216	.0260	.0309	.0361
6	.0008	.0012	.0018	.0026	.0035	.0047	.0061	.0078	.0098	.0120
7	.0001	.0002	.0003	.0005	.0008	.0011	.0015	.0020	.0027	.0034
8	.0000	.0000	.0001	.0001	.0001	.0002	.0003	.0005	.0006	.0009
9	.0000	.0000	.0000	.0000	.0000	.0000	.0001	.0001	.0001	.0002

m

x	2.1	2.2	2.3	2.4	2.5	2.6	2.7	2.8	2.9	3.0
0	.1225	.1108	.1003	.0907	.0821	.0743	.0672	.0608	.0550	.0498
1	.2572	.2438	.2306	.2177	.2052	.1931	.1815	.1703	.1596	.1494
2	.2700	.2681	.2652	.2613	.2565	.2510	.2450	.2384	.2314	.2240
3	.1890	.1966	.2033	.2090	.2138	.2176	.2205	.2225	.2237	.2240
4	.0992	.1082	.1169	.1254	.1336	.1414	.1488	.1557	.1622	.1680
5	.0417	.0476	.0538	.0602	.0668	.0735	.0804	.0872	.0940	.1008
6	.0146	.0174	.0206	.0241	.0278	.0319	.0362	.0407	.0455	.0504
7	.0044	.0055	.0068	.0083	.0099	.0118	.0139	.0163	.0188	.0216
8	.0011	.0015	.0019	.0025	.0031	.0038	.0047	.0057	.0068	.0081
9	.0003	.0004	.0005	.0007	.0009	.0011	.0014	.0018	.0022	.0027
10	.0001	.0001	.0001	.0002	.0002	.0003	.0004	.0005	.0006	.0008
11	.0000	.0000	.0000	.0000	.0000	.0001	.0001	.0001	.0002	.0002
12	.0000	.0000	.0000	.0000	.0000	.0000	.0000	.0000	.0000	.0001

m

x	3.1	3.2	3.3	3.4	3.5	3.6	3.7	3.8	3.9	4.0
0	.0450	.0408	.0369	.0334	.0302	.0273	.0247	.0224	.0202	.0183
1	.1397	.1304	.1217	.1135	.1057	.0984	.0915	.0850	.0789	.0733
2	.2165	.2087	.2008	.1929	.1850	.1771	.1692	.1615	.1539	.1465
3	.2237	.2226	.2209	.2186	.2158	.2125	.2087	.2046	.2001	.1954
4	.1734	.1781	.1823	.1858	.1888	.1912	.1931	.1944	.1951	.1954
5	.1075	.1140	.1203	.1264	.1322	.1377	.1429	.1477	.1522	.1563
6	.0555	.0608	.0662	.0716	.0771	.0826	.0881	.0936	.0989	.1042
7	.0246	.0278	.0312	.0348	.0385	.0425	.0466	.0508	.0551	.0595
8	.0095	.0111	.0129	.0148	.0169	.0191	.0215	.0241	.0269	.0298
9	.0033	.0040	.0047	.0056	.0066	.0076	.0089	.0102	.0116	.0132
10	.0010	.0013	.0016	.0019	.0023	.0028	.0033	.0039	.0045	.0053
11	.0003	.0004	.0005	.0006	.0007	.0009	.0011	.0013	.0016	.0019
12	.0001	.0001	.0001	.0002	.0002	.0003	.0003	.0004	.0005	.0006
13	.0000	.0000	.0000	.0000	.0001	.0001	.0001	.0001	.0002	.0002
14	.0000	.0000	.0000	.0000	.0000	.0000	.0000	.0000	.0000	.0001

(Continued)

Table II (Continued)

x						m				
	4.1	4.2	4.3	4.4	4.5	4.6	4.7	4.8	4.9	5.0
0	.0166	.0150	.0136	.0123	.0111	.0101	.0091	.0082	.0074	.0067
1	.0679	.0630	.0583	.0540	.0500	.0462	.0427	.0395	.0365	.0337
2	.1393	.1323	.1254	.1188	.1125	.1063	.1005	.0948	.0894	.0842
3	.1904	.1852	.1798	.1743	.1687	.1631	.1574	.1517	.1460	.1404
4	.1951	.1944	.1933	.1917	.1898	.1875	.1849	.1820	.1789	.1755
5	.1600	.1633	.1662	.1687	.1708	.1725	.1738	.1747	.1753	.1755
6	.1093	.1143	.1191	.1237	.1281	.1323	.1362	.1398	.1432	.1462
7	.0640	.0686	.0732	.0778	.0824	.0869	.0914	.0959	.1002	.1044
8	.0328	.0360	.0393	.0428	.0463	.0500	.0537	.0575	.0614	.0653
9	.0150	.0168	.0188	.0209	.0232	.0255	.0280	.0307	.0334	.0363
10	.0061	.0071	.0081	.0092	.0104	.0118	.0132	.0147	.0164	.0181
11	.0023	.0027	.0032	.0037	.0043	.0049	.0056	.0064	.0073	.0082
12	.0008	.0009	.0011	.0014	.0016	.0019	.0022	.0026	.0030	.0034
13	.0002	.0003	.0004	.0005	.0006	.0007	.0008	.0009	.0011	.0013
14	.0001	.0001	.0001	.0001	.0002	.0002	.0003	.0003	.0004	.0005
15	.0000	.0000	.0000	.0000	.0001	.0001	.0001	.0001	.0001	.0002

(Continued)

Table II (*Continued*)

x	7.0	8.0	9.0	10
0	.0009	.0003	.0001	.0000
1	.0064	.0027	.0011	.0005
2	.0223	.0107	.0050	.0023
3	.0521	.0286	.0150	.0076
4	.0912	.0573	.0337	.0189
5	.1277	.0916	.0607	.0378
6	.1490	.1221	.0911	.0631
7	.1490	.1396	.1171	.0901
8	.1304	.1396	.1318	.1126
9	.1014	.1241	.1318	.1251
10	.0710	.0993	.1186	.1251
11	.0452	.0722	.0970	.1137
12	.0264	.0481	.0728	.0948
13	.0142	.0296	.0504	.0729
14	.0071	.0169	.0324	.0521
15	.0033	.0090	.0194	.0347
16	.0014	.0045	.0109	.0217
17	.0006	.0021	.0058	.0128
18	.0002	.0009	.0029	.0071
19	.0001	.0004	.0014	.0037
20		.0002	.0006	.0019
21		.0001	.0003	.0009
22			.0001	.0004
23				.0002
24				.0001

Table III   Area $\Phi(z)$ under the normal curve to the left of z

z	.00	.01	.02	.03	.04	.05	.06	.07	.08	.09
0.0	5000	5040	5080	5120	5160	5199	5239	5279	5319	5359
0.1	5398	5438	5478	5517	5557	5596	5636	5675	5714	5753
0.2	5793	5832	5871	5910	5948	5987	6026	6064	6103	6141
0.3	6179	6217	6255	6293	6331	6368	6406	6443	6480	6517
0.4	6554	6591	6628	6664	6700	6736	6772	6808	6844	6879
0.5	6915	6950	6985	7019	7054	7088	7123	7157	7190	7224
0.6	7257	7291	7324	7357	7389	7422	7454	7486	7517	7549
0.7	7580	7611	7642	7673	7704	7734	7764	7794	7823	7852
0.8	7881	7910	7939	7967	7995	8023	8051	8078	8106	8133
0.9	8159	8186	8212	8238	8264	8289	8315	8340	8365	8389
1.0	8413	8438	8461	8485	8508	8531	8554	8577	8599	8621
1.1	8643	8665	8686	8708	8729	8749	8770	8790	8810	8830
1.2	8849	8869	8888	8907	8925	8944	8962	8980	8997	9015
1.3	9032	9049	9066	9082	9099	9115	9131	9147	9162	9177
1.4	9192	9207	9222	9236	9251	9265	9279	9292	9306	9319
1.5	9332	9345	9357	9370	9382	9394	9406	9418	9429	9441
1.6	9452	9463	9474	9484	9495	9505	9515	9525	9535	9545
1.7	9554	9564	9573	9584	9591	9599	9608	9616	9625	9633
1.8	9641	9649	9656	9664	9671	9673	9686	9693	9699	9706
1.9	9713	9719	9726	9732	9738	9744	9750	9756	9761	9767
2.0	9772	9778	9783	9788	9793	9798	9803	9808	9812	9817
2.1	9821	9826	9830	9834	9838	9842	9846	9850	9854	9857
2.2	9861	9864	9868	9871	9875	9878	9881	9884	9887	9890
2.3	9893	9896	9898	9901	9904	9906	9909	9911	9913	9916
2.4	9918	9920	9922	9925	9927	9929	9931	9932	9934	9936
2.5	9938	9940	9941	9943	9945	9946	9948	9949	9950	9952
2.6	9953	9955	9956	9957	9959	9960	9961	9962	9963	9964
2.7	9965	9966	9967	9968	9969	9970	9971	9972	9973	9974
2.8	9974	9975	9976	9977	9977	9978	9979	9979	9980	9981
2.9	9981	9982	9982	9983	9984	9984	9985	9985	9986	9986
3.0	9987	9987	9987	9988	9988	9989	9989	9989	9990	9990
3.1	9990	9991	9991	9991	9992	9992	9992	9992	9993	9993
3.2	9993	9993	9994	9994	9994	9994	9994	9995	9995	9995
3.3	9995	9995	9995	9996	9996	9996	9996	9996	9996	9997
3.4	9997	9997	9997	9997	9997	9997	9997	9997	9997	9998

Table IV Chi-Square Distribution

F \ n	.005	.010	.025	.050	.100	.250	.500	.750	.900	.950	.975	.990	.995
1	.0000393	.000157	.000982	.00393	.0158	.102	.455	1.32	2.71	3.84	5.02	6.63	7.88
2	.0100	.0201	.0506	.103	.211	.575	1.39	2.77	4.61	5.99	7.38	9.21	10.6
3	.0717	.115	.216	.352	.584	1.21	2.37	4.11	6.25	7.81	9.35	11.3	12.8
4	.207	.297	.484	.711	1.06	1.92	3.36	5.39	7.78	9.49	11.1	13.3	14.9
5	.412	.554	.831	1.15	1.61	2.67	4.35	6.63	9.24	11.1	12.8	15.1	16.7
6	.676	.872	1.24	1.64	2.20	3.45	5.35	7.84	10.6	12.6	14.4	16.8	18.5
7	.989	1.24	1.69	2.17	2.83	4.25	6.35	9.04	12.0	14.1	16.0	18.5	20.3
8	1.34	1.65	2.18	2.73	3.49	5.07	7.34	10.2	13.4	15.5	17.5	20.1	22.0
9	1.73	2.09	2.70	3.33	4.17	5.90	8.34	11.4	14.7	16.9	19.0	21.7	23.6
10	2.16	2.56	3.25	3.94	4.87	6.74	9.34	12.5	16.0	18.3	20.5	23.2	25.2
11	2.60	3.05	3.82	4.57	5.58	7.58	10.3	13.7	17.3	19.7	21.9	24.7	26.8
12	3.07	3.57	4.40	5.23	6.30	8.44	11.3	14.8	18.5	21.0	23.3	26.2	28.3
13	3.57	4.11	5.01	5.89	7.04	9.30	12.3	16.0	19.8	22.4	24.7	27.7	29.8
14	4.07	4.66	5.63	6.57	7.79	10.2	13.3	17.1	21.1	23.7	26.1	29.1	31.3
15	4.60	5.23	6.26	7.26	8.55	11.0	14.3	18.2	22.3	25.0	27.5	30.6	32.8
16	5.14	5.81	6.91	7.96	9.31	11.9	15.3	19.4	23.5	26.3	28.8	32.0	34.3
17	5.70	6.41	7.56	8.67	10.1	12.8	16.3	20.5	24.8	27.6	30.2	33.4	35.7
18	6.26	7.01	8.23	9.39	10.9	13.7	17.3	21.6	26.0	28.9	31.5	34.8	37.2
19	6.84	7.63	8.91	10.1	11.7	14.6	18.3	22.7	27.2	30.1	32.9	36.2	38.6
20	7.43	8.26	9.59	10.9	12.4	15.5	19.3	23.8	28.4	31.4	34.2	37.6	40.0
21	8.03	8.90	10.3	11.6	13.2	16.3	20.3	24.9	29.6	32.7	35.5	38.9	41.4
22	8.64	9.54	11.0	12.3	14.0	17.2	21.3	26.0	30.8	33.9	36.8	40.3	42.8
23	9.26	10.2	11.7	13.1	14.8	18.1	22.3	27.1	32.0	35.2	38.1	41.6	44.2
24	9.89	10.9	12.4	13.8	15.7	19.0	23.3	28.2	33.2	36.4	39.4	43.0	45.6
25	10.5	11.5	13.1	14.6	16.5	19.9	24.3	29.3	34.4	37.7	40.6	44.3	46.9
26	11.2	12.2	13.8	15.4	17.3	20.8	25.3	30.4	35.6	38.9	41.9	45.6	48.3
27	11.8	12.9	14.6	16.2	18.1	21.7	26.3	31.5	36.7	40.1	43.2	47.0	49.6
28	12.5	13.6	15.3	16.9	18.9	22.7	27.3	32.6	37.9	41.3	44.5	48.3	51.0
29	13.1	14.3	16.0	17.7	19.8	23.6	28.3	33.7	39.1	42.6	45.7	49.6	52.3
30	13.8	15.0	16.8	18.5	20.6	24.5	29.3	34.8	40.3	43.8	47.0	50.9	53.7

Table V   Student's *t*-Distribution, Percentage Points

n	.60	.75	.90	.95	.975	.99	.995	.9995
1	.325	1.000	3.078	6.314	12.706	31.821	63.657	636.619
2	.289	.816	1.886	2.920	4.303	6.965	9.925	31.598
3	.277	.765	1.638	2.353	3.182	4.541	5.841	12.941
4	.271	.741	1.533	2.132	2.776	3.747	4.604	8.610
5	.267	.727	1.476	2.015	2.571	3.365	4.032	6.859
6	.265	.718	1.440	1.943	2.447	3.143	3.707	5.959
7	.263	.711	1.415	1.895	2.365	2.998	3.499	5.405
8	.262	.706	1.397	1.860	2.306	2.896	3.355	5.041
9	.261	.703	1.383	1.833	2.262	2.821	3.250	4.781
10	.260	.700	1.372	1.812	2.228	2.764	3.169	4.587
11	.260	.697	1.363	1.796	2.201	2.718	3.106	4.437
12	.259	.695	1.356	1.782	2.179	2.681	3.055	4.318
13	.259	.694	1.350	1.771	2.160	2.650	3.012	4.221
14	.258	.692	1.345	1.761	2.145	2.624	2.977	4.140
15	.258	.691	1.341	1.753	2.131	2.602	2.947	4.073
16	.258	.690	1.337	1.746	2.120	2.583	2.921	4.015
17	.257	.689	1.333	1.740	2.110	2.567	2.898	3.965
18	.257	.688	1.330	1.734	2.101	2.552	2.878	3.922
19	.257	.688	1.328	1.729	2.093	2.539	2.861	3.883
20	.257	.687	1.325	1.725	2.086	2.528	2.845	3.850
21	.257	.686	1.323	1.721	2.080	2.518	2.831	3.819
22	.256	.686	1.321	1.717	2.074	2.508	2.819	3.792
23	.256	.685	1.319	1.714	2.069	2.500	2.807	3.767
24	.256	.685	1.318	1.711	2.064	2.492	2.797	3.745
25	.256	.684	1.316	1.708	2.060	2.485	2.787	3.725
26	.256	.684	1.315	1.706	2.056	2.479	2.779	3.707
27	.256	.684	1.314	1.703	2.052	2.473	2.771	3.690
28	.256	.683	1.313	1.701	2.048	2.467	2.763	3.674
29	.256	.683	1.311	1.699	2.045	2.462	2.756	3.659
30	.256	.683	1.310	1.697	2.042	2.457	2.750	3.646
40	.255	.681	1.303	1.684	2.021	2.423	2.704	3.551
60	.254	.679	1.296	1.671	2.000	2.390	2.660	3.460
120	.254	.677	1.289	1.658	1.980	2.358	2.617	3.373
∞	.253	.674	1.282	1.645	1.960	2.326	2.576	3.291

Table VI Percentage Points, F-Distribution

.95

m \ n	1	2	3	4	5	6	7	8	9	10	12	15	20	24	30	40	60	120	∞
1	161.4	199.5	215.7	224.6	230.2	234.0	236.8	238.9	240.5	241.9	243.9	245.9	248.0	249.1	250.1	251.1	252.2	253.3	254.3
2	18.51	19.00	19.16	19.25	19.30	19.33	19.35	19.37	19.38	19.40	19.41	19.43	19.45	19.45	19.46	19.47	19.48	19.49	19.50
3	10.13	9.55	9.28	9.12	9.01	8.94	8.89	8.85	8.81	8.79	8.74	8.70	8.66	8.64	8.62	8.59	8.57	8.55	8.53
4	7.71	6.94	6.59	6.39	6.26	6.16	6.09	6.04	6.00	5.96	5.91	5.86	5.80	5.77	5.75	5.72	5.69	5.66	5.63
5	6.61	5.79	5.41	5.19	5.05	4.95	4.88	4.82	4.77	4.74	4.68	4.62	4.56	4.53	4.50	4.46	4.43	4.40	4.36
6	5.99	5.14	4.76	4.53	4.39	4.28	4.21	4.15	4.10	4.06	4.00	3.94	3.87	3.84	3.81	3.77	3.74	3.70	3.67
7	5.59	4.74	4.35	4.12	3.97	3.87	3.79	3.73	3.68	3.64	3.57	3.51	3.44	3.41	3.38	3.34	3.30	3.27	3.23
8	5.32	4.46	4.07	3.84	3.69	3.58	3.50	3.44	3.39	3.35	3.28	3.22	3.15	3.12	3.08	3.04	3.01	2.97	2.93
9	5.12	4.26	3.86	3.63	3.48	3.37	3.29	3.23	3.18	3.14	3.07	3.01	2.94	2.90	2.86	2.83	2.79	2.75	2.71
10	4.96	4.10	3.71	3.48	3.33	3.22	3.14	3.07	3.02	2.98	2.91	2.85	2.77	2.74	2.70	2.66	2.62	2.58	2.54
11	4.84	3.98	3.59	3.36	3.20	3.09	3.01	2.95	2.90	2.85	2.79	2.72	2.65	2.61	2.57	2.53	2.49	2.45	2.40
12	4.75	3.89	3.49	3.26	3.11	3.00	2.91	2.85	2.80	2.75	2.69	2.62	2.54	2.51	2.47	2.43	2.38	2.34	2.30
13	4.67	3.81	3.41	3.18	3.03	2.92	2.83	2.77	2.71	2.67	2.60	2.53	2.46	2.42	2.38	2.34	2.30	2.25	2.21
14	4.60	3.74	3.34	3.11	2.96	2.85	2.76	2.70	2.65	2.60	2.53	2.46	2.39	2.35	2.31	2.27	2.22	2.18	2.13
15	4.54	3.68	3.29	3.06	2.90	2.79	2.71	2.64	2.59	2.54	2.48	2.40	2.33	2.29	2.25	2.20	2.16	2.11	2.07
16	4.49	3.63	3.24	3.01	2.85	2.74	2.66	2.59	2.54	2.49	2.42	2.35	2.28	2.24	2.19	2.15	2.11	2.06	2.01
17	4.45	3.59	3.20	2.96	2.81	2.70	2.61	2.55	2.49	2.45	2.38	2.31	2.23	2.19	2.15	2.10	2.06	2.01	1.96
18	4.41	3.55	3.16	2.93	2.77	2.66	2.58	2.51	2.46	2.41	2.34	2.27	2.19	2.15	2.11	2.06	2.02	1.97	1.92
19	4.38	3.52	3.13	2.90	2.74	2.63	2.54	2.48	2.42	2.38	2.31	2.23	2.16	2.11	2.07	2.03	1.98	1.93	1.88
20	4.35	3.49	3.10	2.87	2.71	2.60	2.51	2.45	2.39	2.35	2.28	2.20	2.12	2.08	2.04	1.99	1.95	1.90	1.84
21	4.32	3.47	3.07	2.84	2.68	2.57	2.49	2.42	2.37	2.32	2.25	2.18	2.10	2.05	2.01	1.96	1.92	1.87	1.81
22	4.30	3.44	3.05	2.82	2.66	2.55	2.46	2.40	2.34	2.30	2.23	2.15	2.07	2.03	1.98	1.94	1.89	1.84	1.78
23	4.28	3.42	3.03	2.80	2.64	2.53	2.44	2.37	2.32	2.27	2.20	2.13	2.05	2.01	1.96	1.91	1.86	1.81	1.76
24	4.26	3.40	3.01	2.78	2.62	2.51	2.42	2.36	2.30	2.25	2.18	2.11	2.03	1.98	1.94	1.89	1.84	1.79	1.73
25	4.24	3.39	2.99	2.76	2.60	2.49	2.40	2.34	2.28	2.24	2.16	2.09	2.01	1.96	1.92	1.87	1.82	1.77	1.71
26	4.23	3.37	2.98	2.74	2.59	2.47	2.39	2.32	2.27	2.22	2.15	2.07	1.99	1.95	1.90	1.85	1.80	1.75	1.69
27	4.21	3.35	2.96	2.73	2.57	2.46	2.37	2.31	2.25	2.20	2.13	2.06	1.97	1.93	1.88	1.84	1.79	1.73	1.67
28	4.20	3.34	2.95	2.71	2.56	2.45	2.36	2.29	2.24	2.19	2.12	2.04	1.96	1.91	1.87	1.82	1.77	1.71	1.65
29	4.18	3.33	2.93	2.70	2.55	2.43	2.35	2.28	2.22	2.18	2.10	2.03	1.94	1.90	1.85	1.81	1.75	1.70	1.64
30	4.17	3.32	2.92	2.69	2.53	2.42	2.33	2.27	2.21	2.16	2.09	2.01	1.93	1.89	1.84	1.79	1.74	1.68	1.62
40	4.08	3.23	2.84	2.61	2.45	2.34	2.25	2.18	2.12	2.08	2.00	1.92	1.84	1.79	1.74	1.69	1.64	1.58	1.51
60	4.00	3.15	2.76	2.53	2.37	2.25	2.17	2.10	2.04	1.99	1.92	1.84	1.75	1.70	1.65	1.59	1.53	1.47	1.39
120	3.92	3.07	2.68	2.45	2.29	2.17	2.09	2.02	1.96	1.91	1.83	1.75	1.66	1.61	1.55	1.50	1.43	1.35	1.25
∞	3.84	3.00	2.60	2.37	2.21	2.10	2.01	1.94	1.88	1.83	1.75	1.67	1.57	1.52	1.46	1.39	1.32	1.22	1.00

Table VI (Continued)

.975

m \ n	1	2	3	4	5	6	7	8	9	10	12	15	20	24	30	40	60	120	x
1	647.8	799.5	864.2	899.6	921.8	937.1	948.2	956.7	963.3	968.6	976.7	984.9	993.1	997.2	1001	1006	1010	1014	1018
2	38.51	39.00	39.17	39.25	39.30	39.33	39.36	39.37	39.39	39.40	39.41	39.43	39.45	39.46	39.46	39.47	39.48	39.49	39.50
3	17.44	16.04	15.44	15.10	14.88	14.73	14.62	14.54	14.47	14.42	14.34	14.25	14.17	14.12	14.08	14.04	13.99	13.95	13.90
4	12.22	10.65	9.98	9.60	9.36	9.20	9.07	8.98	8.90	8.84	8.75	8.66	8.56	8.51	8.46	8.41	8.36	8.31	8.26
5	10.01	8.43	7.76	7.39	7.15	6.98	6.85	6.76	6.68	6.62	6.52	6.43	6.33	6.28	6.23	6.18	6.12	6.07	6.02
6	8.81	7.26	6.60	6.23	5.99	5.82	5.70	5.60	5.52	5.46	5.37	5.27	5.17	5.12	5.07	5.01	4.96	4.90	4.85
7	8.07	6.54	5.89	5.52	5.29	5.12	4.99	4.90	4.82	4.76	4.67	4.57	4.47	4.42	4.36	4.31	4.25	4.20	4.14
8	7.57	6.06	5.42	5.05	4.82	4.65	4.53	4.43	4.36	4.30	4.20	4.10	4.00	3.95	3.89	3.84	3.78	3.73	3.67
9	7.21	5.71	5.08	4.72	4.48	4.32	4.20	4.10	4.03	3.96	3.87	3.77	3.67	3.61	3.56	3.51	3.45	3.39	3.33
10	6.94	5.46	4.83	4.47	4.24	4.07	3.95	3.85	3.78	3.72	3.62	3.52	3.42	3.37	3.31	3.26	3.20	3.14	3.08
11	6.72	5.26	4.63	4.28	4.04	3.88	3.76	3.66	3.59	3.53	3.43	3.33	3.23	3.17	3.12	3.06	3.00	2.94	2.88
12	6.55	5.10	4.47	4.12	3.89	3.73	3.61	3.51	3.44	3.37	3.28	3.18	3.07	3.02	2.96	2.91	2.85	2.79	2.72
13	6.41	4.97	4.35	4.00	3.77	3.60	3.48	3.39	3.31	3.25	3.15	3.05	2.95	2.89	2.84	2.78	2.72	2.66	2.60
14	6.30	4.86	4.24	3.89	3.66	3.50	3.38	3.29	3.21	3.15	3.05	2.95	2.84	2.79	2.73	2.67	2.61	2.55	2.49
15	6.20	4.77	4.15	3.80	3.58	3.41	3.29	3.20	3.12	3.06	2.96	2.86	2.76	2.70	2.64	2.59	2.52	2.46	2.40
16	6.12	4.69	4.08	3.73	3.50	3.34	3.22	3.12	3.05	2.99	2.89	2.79	2.68	2.63	2.57	2.51	2.45	2.38	2.32
17	6.04	4.62	4.01	3.66	3.44	3.28	3.16	3.06	2.98	2.92	2.82	2.72	2.62	2.56	2.50	2.44	2.38	2.32	2.25
18	5.98	4.56	3.95	3.61	3.38	3.22	3.10	3.01	2.93	2.87	2.77	2.67	2.56	2.50	2.44	2.38	2.32	2.26	2.19
19	5.92	4.51	3.90	3.56	3.33	3.17	3.05	2.96	2.88	2.82	2.72	2.62	2.51	2.45	2.39	2.33	2.27	2.20	2.13
20	5.87	4.46	3.86	3.51	3.29	3.13	3.01	2.91	2.84	2.77	2.68	2.57	2.46	2.41	2.35	2.29	2.22	2.16	2.09
21	5.83	4.42	3.82	3.48	3.25	3.09	2.97	2.87	2.80	2.73	2.64	2.53	2.42	2.37	2.31	2.25	2.18	2.11	2.04
22	5.79	4.38	3.78	3.44	3.22	3.05	2.93	2.84	2.76	2.70	2.60	2.50	2.39	2.33	2.27	2.21	2.14	2.08	2.00
23	5.75	4.35	3.75	3.41	3.18	3.02	2.90	2.81	2.73	2.67	2.57	2.47	2.36	2.30	2.24	2.18	2.11	2.04	1.97
24	5.72	4.32	3.72	3.38	3.15	2.99	2.87	2.78	2.70	2.64	2.54	2.44	2.33	2.27	2.21	2.15	2.08	2.01	1.94
25	5.69	4.29	3.69	3.35	3.13	2.97	2.85	2.75	2.68	2.61	2.51	2.41	2.30	2.24	2.18	2.12	2.05	1.98	1.91
26	5.66	4.27	3.67	3.33	3.10	2.94	2.82	2.73	2.65	2.59	2.49	2.39	2.28	2.22	2.16	2.09	2.03	1.95	1.88
27	5.63	4.24	3.65	3.31	3.08	2.92	2.80	2.71	2.63	2.57	2.47	2.36	2.25	2.19	2.13	2.07	2.00	1.93	1.85
28	5.61	4.22	3.63	3.29	3.06	2.90	2.78	2.69	2.61	2.55	2.45	2.34	2.23	2.17	2.11	2.05	1.98	1.91	1.83
29	5.59	4.20	3.61	3.27	3.04	2.88	2.76	2.67	2.59	2.53	2.43	2.32	2.21	2.15	2.09	2.03	1.96	1.89	1.81
30	5.57	4.18	3.59	3.25	3.03	2.87	2.75	2.65	2.57	2.51	2.41	2.31	2.20	2.14	2.07	2.01	1.94	1.87	1.79
40	5.42	4.05	3.46	3.13	2.90	2.74	2.62	2.53	2.45	2.39	2.29	2.18	2.07	2.01	1.94	1.88	1.80	1.72	1.64
60	5.29	3.93	3.34	3.01	2.79	2.63	2.51	2.41	2.33	2.27	2.17	2.06	1.94	1.88	1.82	1.74	1.67	1.58	1.48
120	5.15	3.80	3.23	2.89	2.67	2.52	2.39	2.30	2.22	2.16	2.05	1.94	1.82	1.76	1.69	1.61	1.53	1.43	1.31
∞	5.02	3.69	3.12	2.79	2.57	2.41	2.29	2.19	2.11	2.05	1.94	1.83	1.71	1.64	1.57	1.48	1.39	1.27	1.00

Table VI (Continued)

.99

n \ m	1	2	3	4	5	6	7	8	9	10	12	15	20	24	30	40	60	120	∞
1	4052	4999.5	5403	5625	5764	5859	5928	5982	6022	6056	6106	6157	6209	6235	6261	6287	6313	6339	6366
2	98.50	99.00	99.17	99.25	99.30	99.33	99.36	99.37	99.39	99.40	99.42	99.43	99.45	99.46	99.47	99.47	99.48	99.49	99.50
3	34.12	30.82	29.46	28.71	28.24	27.91	27.67	27.49	27.35	27.23	27.05	26.87	26.69	26.60	26.50	26.41	26.32	26.22	26.13
4	21.20	18.00	16.69	15.98	15.52	15.21	14.98	14.80	14.66	14.55	14.37	14.20	14.02	13.93	13.84	13.75	13.65	13.56	13.46
5	16.26	13.27	12.06	11.39	10.97	10.67	10.46	10.29	10.16	10.05	9.89	9.72	9.55	9.47	9.38	9.29	9.20	9.11	9.02
6	13.75	10.92	9.78	9.15	8.75	8.47	8.26	8.10	7.98	7.87	7.72	7.56	7.40	7.31	7.23	7.14	7.06	6.97	6.88
7	12.25	9.55	8.45	7.85	7.46	7.19	6.99	6.84	6.72	6.62	6.47	6.31	6.16	6.07	5.99	5.91	5.82	5.74	5.65
8	11.26	8.65	7.59	7.01	6.63	6.37	6.18	6.03	5.91	5.81	5.67	5.52	5.36	5.28	5.20	5.12	5.03	4.95	4.86
9	10.56	8.02	6.99	6.42	6.06	5.80	5.61	5.47	5.35	5.26	5.11	4.96	4.81	4.73	4.65	4.57	4.48	4.40	4.31
10	10.04	7.56	6.55	5.99	5.64	5.39	5.20	5.06	4.94	4.85	4.71	4.56	4.41	4.33	4.25	4.17	4.08	4.00	3.91
11	9.65	7.21	6.22	5.67	5.32	5.07	4.89	4.74	4.63	4.54	4.40	4.25	4.10	4.02	3.94	3.86	3.78	3.69	3.60
12	9.33	6.93	5.95	5.41	5.06	4.82	4.64	4.50	4.39	4.30	4.16	4.01	3.86	3.78	3.70	3.62	3.54	3.45	3.36
13	9.07	6.70	5.74	5.21	4.86	4.62	4.44	4.30	4.19	4.10	3.96	3.82	3.66	3.59	3.51	3.43	3.34	3.25	3.17
14	8.86	6.51	5.56	5.04	4.69	4.46	4.28	4.14	4.03	3.94	3.80	3.66	3.51	3.43	3.35	3.27	3.18	3.09	3.00
15	8.68	6.36	5.42	4.89	4.56	4.32	4.14	4.00	3.89	3.80	3.67	3.52	3.37	3.29	3.21	3.13	3.05	2.96	2.87
16	8.53	6.23	5.29	4.77	4.44	4.20	4.03	3.89	3.78	3.69	3.55	3.41	3.26	3.18	3.10	3.02	2.93	2.84	2.75
17	8.40	6.11	5.18	4.67	4.34	4.10	3.93	3.79	3.68	3.59	3.46	3.31	3.16	3.08	3.00	2.92	2.83	2.75	2.65
18	8.29	6.01	5.09	4.58	4.25	4.01	3.84	3.71	3.60	3.51	3.37	3.23	3.08	3.00	2.92	2.84	2.75	2.66	2.57
19	8.18	5.93	5.01	4.50	4.17	3.94	3.77	3.63	3.52	3.43	3.30	3.15	3.00	2.92	2.84	2.76	2.67	2.58	2.49
20	8.10	5.85	4.94	4.43	4.10	3.87	3.70	3.56	3.46	3.37	3.23	3.09	2.94	2.86	2.78	2.69	2.61	2.52	2.42
21	8.02	5.78	4.87	4.37	4.04	3.81	3.64	3.51	3.40	3.31	3.17	3.03	2.80	2.80	2.72	2.64	2.55	2.46	2.36
22	7.95	5.72	4.82	4.31	3.99	3.76	3.59	3.45	3.35	3.26	3.12	2.98	2.83	2.75	2.67	2.58	2.50	2.40	2.31
23	7.88	5.66	4.76	4.26	3.94	3.71	3.54	3.41	3.30	3.21	3.07	2.93	2.78	2.70	2.62	2.54	2.45	2.35	2.26
24	7.82	5.61	4.72	4.22	3.90	3.67	3.50	3.36	3.26	3.17	3.03	2.89	2.74	2.66	2.58	2.49	2.40	2.31	2.21
25	7.77	5.57	4.68	4.18	3.85	3.63	3.46	3.32	3.22	3.13	2.99	2.85	2.70	2.62	2.54	2.45	2.36	2.27	2.17
26	7.72	5.53	4.64	4.14	3.82	3.59	3.42	3.29	3.18	3.09	2.96	2.81	2.66	2.58	2.50	2.42	2.33	2.23	2.13
27	7.68	5.49	4.60	4.11	3.78	3.56	3.39	3.26	3.15	3.06	2.93	2.78	2.63	2.55	2.47	2.38	2.29	2.20	2.10
28	7.64	5.45	4.57	4.07	3.75	3.53	3.36	3.23	3.12	3.03	2.90	2.75	2.60	2.52	2.44	2.35	2.26	2.17	2.06
29	7.60	5.42	4.54	4.04	3.73	3.50	3.33	3.20	3.09	3.00	2.87	2.73	2.57	2.49	2.41	2.33	2.23	2.14	2.03
30	7.56	5.39	4.51	4.02	3.70	3.47	3.30	3.17	3.07	2.98	2.84	2.70	2.55	2.47	2.39	2.30	2.21	2.11	2.01
40	7.31	5.18	4.31	3.83	3.51	3.29	3.12	2.99	2.89	2.80	2.66	2.52	2.37	2.29	2.20	2.11	2.02	1.92	1.80
60	7.08	4.98	4.13	3.65	3.34	3.12	2.95	2.82	2.72	2.63	2.50	2.35	2.20	2.12	2.03	1.94	1.84	1.73	1.60
120	6.85	4.79	3.95	3.48	3.17	2.96	2.79	2.66	2.56	2.47	2.34	2.19	2.03	1.95	1.86	1.76	1.66	1.53	1.38
∞	6.63	4.61	3.78	3.32	3.02	2.80	2.64	2.51	2.41	2.32	2.18	2.04	1.88	1.79	1.70	1.59	1.47	1.32	1.00

Table VII   Random Numbers

95673	19543	88787	24433	65904	33551	74156	48065	38021	46480
74699	83908	01636	10964	49671	19779	51751	47927	17071	20272
70964	46447	35542	21274	23133	59964	15201	69740	97523	40853
43881	95137	67017	47898	03066	94299	83206	46872	92561	45597
17298	83805	44977	07091	60002	15728	46555	64622	07046	07096
94358	44992	10243	83217	00432	35232	43151	94761	72008	21788
84733	34751	53178	70870	11553	67251	72441	22549	63872	86555
27231	64931	47501	81075	34915	40922	75190	51229	71198	09826
74058	01486	97119	26790	78232	19669	67726	23840	31358	76326
02073	85738	88152	18375	67759	87106	28086	98353	07890	82849
13425	94256	75965	36623	87373	20687	42116	01786	03205	72838
69336	60414	41509	80611	26156	21857	51306	12693	55495	31218
99829	26683	48063	98001	36302	41358	74959	68716	92629	91443
65332	55608	21062	25375	65793	36247	15535	38840	22463	49372
12611	96265	72622	40180	44743	76773	31743	72756	32350	91186
22350	84735	71650	27600	36726	37979	68523	79521	47438	24225
70550	29182	49634	37272	21755	11056	05818	30061	52264	47081
40264	47942	97587	28802	13850	51816	98508	04436	30403	40009
40820	58402	47756	42273	00206	19714	66010	12570	51211	91832
74458	35792	17575	91216	99362	29454	66163	11517	65206	59509
46445	44109	64962	08813	66088	26043	19323	96816	68625	88192
26825	53912	12377	65681	99184	72798	12001	76178	82032	62195
37787	17836	22366	09090	15502	64203	79524	07225	29282	80327
01850	09763	81675	18728	41239	78901	75171	92338	74263	71451
45026	93368	66406	40339	78710	24857	03402	44700	25633	85993
05365	91765	28390	57363	69137	90608	87436	50299	36386	42989
92200	48616	30259	57543	54630	11377	36038	03942	95074	40907
26920	64882	73326	11730	75555	71631	56031	39323	62212	20701
76616	12670	56849	78288	22511	06086	57123	39359	61089	81535
45672	12722	94260	91843	97630	50406	36517	43575	01428	10835
04799	31526	45819	46640	94612	33026	75223	05604	76273	64496
41793	27126	24953	87463	01581	92186	52904	70915	02432	53242
46757	36369	96190	05797	58470	16021	38645	81241	52793	97916
55272	19065	49227	23226	38656	11896	21525	97572	47983	50040
05466	33569	88498	35202	67652	02301	30448	72718	19724	41239

Top panel:

53196	90907	65415	45497	10395	29588
79152	73727	71719	50386	15205	41576
13092	38945	97165	11689	09036	71932
20913	33536	92467	33345	65474	13767
15122	23827	42564	80293	42518	01842
87214	56388	16245	39322	41467	43511
20573	20679	17218	44961	49882	32796
18973	66963	70019	53685	26709	84100
56655	98385	49705	05481	40592	49747
76267	24651	00794	75032	10105	49700
39480	72015	95042	93319	09869	04823
08514	50133	15744	15517	13607	92836
91319	93111	07487	51463	21486	59773
90133	07000	09747	71216	65398	90824
97593	18647	26251	26321	44325	34845
91284	20263	16039	94491	33767	73915
95514	56543	06636	61291	67504	57205
04996	82256	47375	87507	05112	88489
18378	32077	36523	30843	07057	78326
92222	16704	00197	51267	33224	40276
65214	33885	82939	50723	88987	69761
30375	64836	18459	08235	67650	72390
64657	42013	12753	03028	24224	24918
58653	38063	25072	48698	88083	48040
43743	82050	78412	79456	95032	10984
42539	91907	25694	37097	39566	24043
66126	54146	67213	52234	48381	89442
84643	40792	47562	95494	62087	18064
36287	03046	87136	36057	93913	70080
27504	21121	94711	11807	80882	48359
97247	24875	26259	67622	14657	80354
60650	76219	02772	48651	66449	89213
95493	16106	12783	32748	83533	25440
83008	20544	41665	99439	70606	28974
78489	12626	60661	53733	70062	14289

Bottom panel:

49251	81598	72663	10875
71016	94188	14369	43123
34010	87907	66345	89062
19562	90990	30889	88923
12974	51394	57811	38840
13568	68468	56238	25776
80928	90362	67884	49942
62212	81543	93254	78969
27143	68153	01084	45781
24397	56650	66567	80216
30157	99348	80868	35088
15161	30403	04924	93707
73005	64826	79875	88689
38834	40095	04605	96708
97243	94836	17184	12353
31692	51607	89056	74472
82997	58320	04852	52595
05043	40582	46051	60261
75781	38768	70475	00601
21033	15175	30741	45814
99092	60991	12571	71753
07204	93373	85112	29610
88859	97254	07771	21393
30497	91407	72900	15699
09726	18075	45852	54968
95330	01985	24128	60514
09760	32388	05601	49923
01534	81967	15337	95831
11234	59350	48368	57195
71056	48762	80221	59683
34208	05374	60304	43178
47132	62839	82198	92445
55685	93302	43019	45861
17803	18184	10510	27159
55045	17219	66737	59080

Table VII (*Continued*)

01923	33647	98442	59293	83318	33425	76412	87062	01295	11083
07202	76476	71888	54845	17468	41964	68694	59662	55905	26898
68825	68242	95750	11033	58634	78411	08523	19313	29327	47526
68525	06496	17446	41378	32368	82019	66101	56733	43308	82641
80819	33515	97373	43064	16221	99697	37951	07947	12935	49391
64200	96929	26044	49283	56545	67200	21325	85056	51345	06309
30156	29121	75874	42399	41121	90643	19585	06364	47203	19679
50467	14282	89098	66717	14753	73356	47781	34156	82842	00121
53764	83212	26675	64184	64455	29023	03181	13674	08838	83829
81727	35572	95469	36825	81882	95083	68323	14965	34166	32351
30807	55558	96026	97398	21723	86560	52617	07771	61886	48234
75104	23682	78756	72728	85940	57290	75507	78715	01426	02310
06180	62724	36835	80288	25075	32609	33312	21348	87710	55457
22098	34834	66117	36252	82717	50585	43639	79999	07414	84003
13173	64783	20984	11929	18849	26211	77375	49561	96747	67007
75273	36108	55265	15653	82270	99216	27805	60088	06056	97377
89849	65756	44454	04602	14292	74458	57777	35934	05160	26359
91108	43562	18883	16569	49599	73871	67101	12054	56492	15981
51843	01542	17881	12954	94913	39583	94969	61146	35907	72184
02644	23564	85464	62947	92571	89377	85004	84654	20465	86212
38608	83374	74032	62183	08740	05279	30455	31032	71512	16476
43164	28909	88624	14992	85359	10193	32491	14769	63694	92640
80933	52950	45646	36636	05085	28053	27596	54873	68476	65823
67690	96766	69250	19344	47855	43489	77479	62418	54079	40069
68579	17014	25362	15114	30982	27250	29052	71115	83369	46776
46353	39733	44677	50133	26623	15979	10651	04263	34087	67005
30039	09532	52215	09164	20930	88230	43403	63230	83525	93550
89200	92772	42195	91634	39272	46462	76835	27755	03151	75692
58118	57942	14807	68214	76093	47484	24468	91764	52907	16675
97230	33027	70166	43232	98802	70715	30216	35586	18909	79658
75569	40225	17892	88888	01568	90289	82347	74237	06083	29023
26618	17921	46235	08527	81088	71103	28867	48898	11765	20589
21391	95668	15669	06585	07408	94593	74192	16937	30915	69683
27236	78121	37636	37596	36916	19026	83895	37621	38781	34507
64567	03291	55990	07279	31596	50559	31313	18888	66235	43465

50375	34061	86395	79626	05026	51164	23098	59665	52597	56451
88089	03283	12554	24989	43509	67955	03735	86889	86277	06435
59112	26102	68353	06053	12603	90508	11316	75692	34749	87629
12474	66011	44885	94478	38566	21045	82523	35402	25612	62630
90484	58184	42161	38435	96744	61538	13021	54147	14388	30496
57547	52673	24130	77619	81861	42763	59069	25009	14725	33579
39797	27602	56402	53611	56680	82374	47413	14069	13527	01958
42855	74811	94877	32342	75691	75512	90283	73490	14789	94230
16868	45219	08345	02402	55789	92892	34715	60781	73791	82930
45414	56595	36151	26808	99662	34688	24116	85249	48272	26688
98889	26797	80541	25881	76837	81704	66596	18122	77787	88508
14822	73860	07541	04333	41893	24404	96965	46622	76266	73091
33845	48523	82848	09751	69494	15664	07836	43250	11078	28256
55067	40413	35370	98082	91804	02184	11189	28913	16663	38476
86131	39325	26626	85304	44087	67037	72628	40755	62226	68424
28756	87618	00781	37579	61867	48251	46943	61543	42343	98703
76657	98868	65010	39343	97240	62107	71035	37373	55414	94309
57193	01772	41594	63353	86262	64660	11696	07440	93454	58572
06385	29257	41642	87040	87970	05607	32126	17875	96175	11150
89797	95386	68790	24040	20909	40305	35493	24822	22344	09095
84006	34817	23061	83258	04742	48610	94168	59039	52179	25656
07986	72678	16554	92420	83521	18540	45819	99214	13509	26480
84387	22967	53018	84589	98830	70090	86620	37444	93863	68686
87840	74766	35877	20896	42826	19807	83228	61427	99636	67563
88675	65449	09582	83371	57878	17567	29258	39952	55896	69821
12054	98757	99170	23056	29309	75640	29123	80829	48533	67423
87001	44944	91536	30186	06647	64319	76847	35269	55755	83992
61875	67634	49064	47964	67275	56073	11594	29891	51162	86276
44660	01229	74140	47725	90783	84239	79045	19065	11766	84993
48611	36739	17009	58754	19081	41171	75088	96714	00713	47112
59552	78041	06587	48476	27370	29730	88393	66475	56061	88242
62935	52864	38961	00910	86829	99746	02686	56438	64551	89416
21507	77663	64810	80323	67246	47898	71120	41330	52403	83716
80891	52275	51237	19891	67753	07891	94253	60222	26110	11708
26421	60075	59796	56401	95486	67147	36635	75026	91863	41944

Table VII (Continued)

03021	43361	52536	10373	07104	86742	72528	64417	32133	66567
42105	81529	16306	67532	94275	45562	85060	44953	50504	78212
19500	57358	62402	42875	94389	66948	33232	60277	31451	95468
01217	39948	55572	50491	04927	99659	00034	25870	72949	97012
03477	74227	05252	37305	81236	22503	96029	55651	27214	40330
09198	25062	23367	32032	63893	91226	74337	44930	39311	83856
53786	97117	65663	66148	02659	67344	23653	92524	38192	86917
72718	84299	32088	74273	83368	90479	32225	93842	73824	26769
24912	20107	14564	42788	05152	65014	08765	03394	49626	03457
72348	93453	71317	91259	36486	76705	80510	72505	72285	00509
61721	03748	38841	69166	01752	51219	78247	70734	10493	25093
98963	01854	20841	52865	31500	09283	07363	71612	77927	61691
23932	43062	33464	15924	24387	75728	31074	53262	70995	42071
15671	70400	88477	85686	91199	11089	56188	53946	30159	59284
73782	89043	22541	79933	57728	87000	97222	97795	16128	55709
83102	95333	72207	71472	52329	37785	52626	28600	13567	45016
46384	82071	00261	22133	43136	74270	98330	82464	44016	03638
58386	47279	85564	56751	02884	47595	53203	37306	69859	22351
31677	34913	92024	95851	73195	55542	52537	17487	95617	64009
90341	27029	35388	10015	54091	25240	28945	51638	55394	00151
53694	56321	51780	85766	24303	91130	08054	93015	83580	78969
14321	34850	57577	81552	88279	30410	36607	00387	97499	10178
05124	43526	80602	08431	39622	90293	58612	37471	28521	98765
04292	67049	37226	77648	62162	92421	33333	76923	38256	78580
09273	84395	97382	42955	25730	55790	91405	34801	64616	44550
15183	69840	80256	57175	06706	71873	79425	03789	29375	04018
88721	47483	15683	85491	46116	03661	47229	63104	29262	61523
43140	44368	16104	46863	53687	53452	78524	02933	56326	83043
91908	49963	85656	41706	18596	42334	71506	69013	64856	48793
96335	20856	03851	47753	99980	24449	39005	58497	03063	68636
87866	14100	55056	01560	58002	26742	43174	30123	39842	37466
78236	79903	57595	30611	49302	87218	77848	38069	70385	36871
36780	08641	20571	13793	09612	27066	20227	50363	44940	38920
18366	06852	71154	94608	70392	04956	21448	09296	19308	61598
45576	91644	74141	23815	82267	28271	81891	31523	20613	46854

89567	21054	47503	25472	75618	13071	81465	31184	23883	03637
23892	65090	87940	33934	31962	59572	77446	83552	31265	38746
31487	44209	43610	01733	05897	66428	34733	44085	58254	37004
46686	41250	54473	98489	59263	22269	45986	60634	51370	17736
71833	75595	31885	08056	46018	99133	90402	28771	43275	34353
72439	98123	65087	40454	51519	56816	30889	15487	81497	75570
77297	37843	69199	76739	18081	58820	85890	87243	84176	35053
33099	33358	55914	34239	47499	14384	13438	39943	86175	28198
95151	61859	09679	83413	91905	36927	18890	57426	28948	02750
74862	50549	49476	62215	87910	43389	28230	64566	36542	10223
10078	82651	08775	29037	36499	33125	98741	71329	21150	07281
56261	21774	62041	84210	04967	23023	88794	40799	96828	01518
06678	84070	69595	48871	86757	92546	81373	54006	67649	56869
34845	41877	95482	84017	32677	11746	69577	07078	12507	58072
80515	53345	11650	03594	64422	97801	26586	82586	54104	26852
32368	38947	68635	82913	60872	11703	40814	67598	28132	64838
80379	42553	16780	91837	61799	70339	45738	71862	56776	15540
72433	90816	97471	61115	17983	26639	45691	24106	42228	00493
87540	87245	61508	31508	36613	00978	03175	34398	14783	88859
92059	84967	49256	33531	79632	66030	15062	01204	23499	32555
12060	41495	77722	23111	04569	70487	87652	27892	93641	51367
39434	85821	57004	00302	24395	67592	50929	27203	56495	44117
49351	81841	20972	12359	52778	14954	28085	90846	75750	72350
87831	38461	72475	26174	54460	45191	88378	78906	57966	61603
25764	17666	45682	74833	92963	48400	13993	70089	69582	82222

Introduction to Statistical Methods

Table VIII   Percentage Points, Distribution of the Correlation Coefficient, when $\rho = 0$

v	$\alpha = 0.05$	0.025	0.01	0.005	0.0025	0.0005	v	$\alpha = 0.05$	0.025	0.01	0.005	0.0025	0.0005
	$2\alpha = 0.1$	0.05	0.02	0.01	0.005	0.001		$2\alpha = 0.1$	0.05	0.02	0.01	0.005	0.001
1	0.9877	0.9^2692	0.9^3507	0.9^3877	0.9^4692	0.9^5877	16	0.400	0.468	0.543	0.590	0.631	0.708
2	.9000	.9500	.9800	$.9^2000$	$.9^2500$	$.9^3000$	17	.389	.456	.529	.575	.616	.693
3	.805	.878	.9343	.9587	.9740	$.9^2114$	18	.378	.444	.516	.561	.602	.679
4	.729	.811	.882	.9172	.9417	.9741	19	.369	.433	.503	.549	.589	.665
5	.669	.754	.833	.875	.9056	.9509	20	.360	.423	.492	.537	.576	.652
6	0.621	0.707	0.789	0.834	0.870	0.9249	25	0.323	0.381	0.445	0.487	0.524	0.597
7	.582	.666	.750	.798	.836	.898	30	.296	.349	.409	.449	.484	.554
8	.549	.632	.715	.765	.805	.872	35	.275	.325	.381	.418	.452	.519
9	.521	.602	.685	.735	.776	.847	40	.257	.304	.358	.393	.425	.490
10	.497	.576	.558	.708	.750	.823	45	.243	.288	.338	.372	.403	.465
11	0.476	0.553	0.634	0.684	0.726	0.801	50	0.231	0.273	0.322	0.354	0.354	0.443
12	.457	.532	.612	.661	.703	.780	60	.211	.250	.295	.325	.352	.408
13	.441	.514	.592	.641	.683	.760	70	.195	.232	.274	.302	.327	.380
14	.426	.497	.574	.623	.664	.742	80	.183	.217	.257	.283·	.307	.357
15	.412	.482	.558	.606	.647	.725	90	.173	.205	.242	.267	.290	.338
							100	.164	.195	.230	.254	.276	.321

Note: α is the upper-tail area of the distribution of r appropriate for use in a single-tail test. For a two-tail test, 2α must be used. If r is calculated from n paired observations, enter the table with $v = n - 2$.

Table IX Upper 5 Percent Points of the Studentized Range

The entries are $q_{.05}$, where $P(q < q_{.05}) = .95$

v \ n	2	3	4	5	6	7	8	9	10
1	17.97	26.98	32.82	37.08	40.41	43.12	45.40	47.36	49.07
2	6.08	8.33	9.80	10.88	11.74	12.44	13.03	13.54	13.99
3	4.50	5.91	6.82	7.50	8.04	8.48	8.85	9.18	9.46
4	3.93	5.04	5.76	5.29	6.71	7.05	7.35	7.60	7.83
5	3.64	4.60	5.22	5.67	6.03	6.33	6.58	6.80	6.99
6	3.46	4.34	4.90	5.30	5.63	5.90	6.12	6.32	6.49
7	3.34	4.16	4.68	5.06	5.36	5.61	5.82	6.00	6.16
8	3.26	4.04	4.53	4.89	5.17	5.40	5.60	5.77	5.92
9	3.20	3.95	4.41	4.76	5.02	5.24	5.43	5.59	5.74
10	3.15	3.88	4.33	4.65	4.91	5.12	5.30	5.46	5.60
11	3.11	3.82	4.26	4.57	4.82	5.03	5.20	5.25	5.49
12	3.08	3.77	4.20	4.51	4.75	4.95	5.12	5.27	5.39
13	3.06	3.73	4.15	4.45	4.69	4.88	5.05	5.19	5.32
14	3.03	3.70	4.11	4.41	4.64	4.83	4.99	5.13	5.25
15	3.01	3.67	4.08	4.37	4.59	4.78	4.94	5.08	5.20
16	3.00	3.65	4.05	4.33	4.56	4.74	4.90	5.03	5.15
17	2.98	3.63	4.02	4.30	4.52	4.70	4.86	4.99	5.11
18	2.97	3.61	4.00	4.28	4.49	4.67	4.82	4.96	5.07
19	2.96	3.59	3.98	4.25	4.47	4.65	4.79	4.92	5.04
20	2.95	3.58	3.96	4.23	4.45	4.62	4.77	490	5.01
24	2.92	3.53	3.90	4.17	4.37	4.54	4.68	4.81	4.92
30	2.89	3.49	3.85	4.10	4.30	4.46	4.60	4.72	4.82
40	2.86	3.44	3.79	4.04	4.23	4.39	4.52	4.63	4.73
60	2.86	3.40	3.74	3.98	4.16	4.31	4.44	4.55	4.65
120	2.80	3.36	3.68	3.92	4.10	4.24	4.36	4.47	4.56
∞	2.77	3.31	3.63	3.86	4.03	4.17	4.29	4.39	4.47

v \ n	11	12	13	14	15	16	17	18	19	20
1	50.59	51.96	53.20	54.33	55.36	56.32	57.22	58.04	58.83	59.56
2	14.39	14.75	15.08	15.38	15.65	15.91	16.14	16.37	16.57	16.77
3	9.72	9.95	10.15	10.35	10.53	10.69	10.84	10.98	11.11	11.24
4	8.03	8.21	8.37	8.52	8.66	8.79	8.91	9.03	9.13	9.23
5	7.17	7.32	7.47	7.60	7.72	7.83	7.93	8.03	8.12	8.21
6	6.65	6.79	6.92	7.03	7.14	7.24	7.34	7.43	7.51	7.59
7	6.30	6.43	6.55	6.66	6.76	6.85	6.94	7.02	7.10	7.17
8	6.05	6.18	6.29	6.39	6.48	6.57	6.65	6.73	6.80	6.87
9	5.87	5.98	6.09	6.19	6.28	6.36	6.44	6.51	6.58	6.64
10	5.72	5.83	5.93	6.03	6.11	6.19	6.27	6.34	6.40	6.47
11	5.61	5.71	5.81	5.90	5.98	6.06	6.13	6.20	6.27	6.33
12	5.51	5.61	5.71	5.80	5.88	5.95	6.02	6.09	6.15	6.21
13	5.43	5.53	5.63	5.71	5.79	5.86	5.93	5.99	6.05	6.11
14	5.36	5.46	5.55	5.64	5.71	5.79	5.85	5.91	5.97	6.03
15	5.31	5.40	5.49	5.57	5.65	5.72	5.78	5.85	5.90	5.96
16	5.26	5.35	5.44	5.52	5.59	5.66	5.73	5.79	5.84	5.90
17	5.21	5.31	5.39	5.47	5.54	5.61	5.67	5.73	5.79	5.84
18	5.17	5.27	5.35	5.43	5.50	5.57	5.63	5.69	5.74	5.79
19	5.14	5.23	5.31	5.39	5.46	5.53	5.59	5.65	5.70	5.75
20	5.11	5.20	5.28	5.36	5.43	5.49	5.55	5.61	5.66	5.71
24	5.01	5.10	5.18	5.25	5.32	5.38	5.44	5.49	5.55	5.59
30	4.92	5.00	5.08	5.15	5.21	5.27	5.33	5.38	5.43	5.47
40	4.82	4.90	4.98	5.04	5.11	5.16	5.22	5.27	5.31	5.36
60	4.73	4.81	4.88	4.94	5.00	5.06	5.11	5.15	5.20	5.24
120	4.64	4.71	4.78	4.84	4.90	4.95	5.00	5.04	5.09	5.13
∞	4.55	4.62	4.68	4.74	4.80	4.85	4.89	4.93	4.97	5.01

Table IX (*Continued*) **Upper 1 Percent Points of the Studentized Range**

The entries are $q_{.01}$, where $P(q < q_{.01}) = .99$

v \ n	2	3	4	5	6	7	8	9	10
1	90.03	135.0	164.3	185.6	202.2	215.2	227.2	237.0	245.6
2	14.04	19.02	22.29	24.72	26.63	28.20	29.53	30.68	31.69
3	8.26	10.62	12.17	13.23	14.24	15.00	154	16.20	16.69
4	6.51	8.12	9.17	9.96	10.58	11.10	11.55	11.93	12.27
5	5.70	6.98	7.80	8.42	8.91	9.32	9.67	9.97	10.24
6	5.24	6.33	7.03	7.56	7.97	8.32	8.61	8.87	9.10
7	4.95	5.92	6.54	7.01	7.37	7.68	7.94	8.17	8.37
8	4.75	5.64	6.20	6.62	6.96	7.24	7.47	7.68	7.86
9	4.60	5.43	5.96	6.35	6.66	6.91	7.13	7.33	7.49
10	4.48	5.27	5.77	6.14	6.43	6.67	6.87	7.05	7.21
11	4.39	5.15	5.62	5.97	6.25	6.48	6.67	6.84	6.99
12	4.32	5.05	5.50	5.84	6.10	6.32	6.51	6.67	6.81
13	4.26	496	5.40	5.73	5.98	6.19	6.37	6.53	6.67
14	4.21	4.89	5.32	5.63	5.88	6.08	6.26	6.41	6.54
15	4.17	4.84	5.25	5.56	5.80	5.99	6.16	6.31	6.44
16	4.13	4.79	5.19	5.49	5.72	5.92	6.08	6.22	6.35
17	4.10	4.74	5.14	5.43	5.66	5.85	6.01	6.15	6.27
18	4.07	4.70	5.09	5.38	5.60	5.79	5.94	6.08	6.20
19	4.05	4.67	5.05	5.33	5.55	5.73	5.89	6.02	6.14
20	4.02	4.64	5.02	5.29	5.51	5.69	5.84	5.97	6.09
24	3.96	4.55	4.91	5.17	5.37	5.54	5.69	5.81	5.92
30	3.89	4.45	4.80	5.05	5.24	5.40	5.54	5.65	5.76
40	3.82	4.37	4.70	4.93	5.11	5.26	5.39	5.50	5.60
60	3.76	4.28	4.59	4.82	4.99	5.13	5.25	5.36	5.45
120	3.70	4.20	4.50	4.71	4.87	5.01	5.12	5.21	5.30
∞	3.64	4.12	4.40	4.60	4.76	4.88	4.99	5.08	5.16

v \ n	11	12	13	14	15	16	17	18	19	20
1	253.2	260.0	266.2	271.8	277.0	281.8	286.3	290.4	294.3	298.0
2	35.59	33.40	34.13	34.81	35.43	36.00	36.53	37.03	37.50	37.95
3	17.13	17.53	17.89	18.22	18.52	18.81	19.07	19.32	19.55	19.77
4	12.57	12.84	13.09	13.32	13.53	13.73	13.91	14.08	14.24	14.40
5	10.48	10.70	10.89	11.08	11.24	11.40	11.55	11.68	11.81	11.93
6	9.30	9.48	9.65	9.81	9.95	10.08	10.21	10.32	10.43	10.54
7	8.55	8.71	8.86	9.00	9.12	9.24	9.35	9.46	9.55	9.65
8	8.03	8.18	8.31	8.44	8.55	8.66	8.76	8.85	8.94	9.03
9	7.65	7.78	7.91	8.03	8.13	8.23	8.33	8.41	8.49	8.57
10	7.36	7.49	7.60	7.71	7.81	7.91	7.99	8.08	8.15	8.23
11	7.13	7.25	7.36	7.46	7.56	7.65	7.73	7.81	7.88	7.95
12	6.94	7.06	7.17	7.26	7.36	7.44	7.52	7.59	7.66	7.73
13	6.79	6.90	7.01	7.10	7.19	7.27	7.35	7.42	7.48	7.55
14	6.66	6.77	6.87	6.96	7.05	7.13	7.20	7.27	7.33	7.39
15	6.55	6.66	6.76	6.84	6.93	7.00	7.07	7.14	7.20	7.26
16	6.46	6.56	6.66	6.74	6.82	6.90	6.97	7.03	7.09	7.15
17	6.38	6.48	6.57	6.66	6.73	6.81	6.87	6.94	7.00	7.05
18	6.31	6.41	6.50	6.58	6.65	6.73	6.79	6.85	6.91	6.97
19	6.25	6.34	6.43	6.51	6.58	6.65	6.72	6.78	6.84	6.89
20	6.19	6.28	6.37	6.45	6.52	6.59	6.65	6.71	6.77	6.82
24	6.02	6.11	6.19	6.26	6.33	6.39	6.45	6.51	6.56	6.61
30	5.85	5.93	6.01	6.08	6.14	6.20	6.26	6.31	6.36	6.41
40	5.69	5.76	5.83	5.90	5.96	6.02	6.07	6.12	6.16	6.21
60	5.53	5.60	5.67	5.73	5.78	5.84	5.89	5.93	5.97	6.01
120	5.37	5.44	5.50	5.56	5.61	5.66	5.71	5.75	5.79	5.83
∞	5.23	5.29	5.35	5.40	5.45	5.49	5.54	5.57	5.61	5.65

Table X   Critical Values of Spearman's Rank Correlation Coefficient

n	$\gamma = 0.10$	$\gamma = 0.05$	$\gamma = 0.02$	$\gamma = 0.01$
5	0.900	—	—	—
6	0.829	0.886	0.943	—
7	0.714	0.786	0.893	—
8	0.643	0.738	0.833	0.881
9	0.600	0.683	0.783	0.833
10	0.564	0.648	0.745	0.794
11	0.523	0.623	0.736	0.818
12	0.497	0.591	0.703	0.780
13	0.475	0.566	0.673	0.745
14	0.457	0.545	0.646	0.716
15	0.441	0.525	0.623	0.689
16	0.425	0.507	0.601	0.666
17	0.412	0.490	0.582	0.645
18	0.399	0.476	0.564	0.625
19	0.388	0.462	0.549	0.608
20	0.377	0.450	0.534	0.591
21	0.368	0.438	0.521	0.576
22	0.359	0.428	0.508	0.562
23	0.351	0.418	0.496	0.549
24	0.343	0.409	0.485	0.537
25	0.336	0.400	0.475	0.526
26	0.329	0.392	0.465	0.515
27	0.323	0.385	0.456	0.505
28	0.317	0.377	0.448	0.496
29	0.311	0.370	0.440	0.487
30	0.305	0.364	0.432	0.478

Table XI d-Factors for Sign Test and Confidence Intervals for the Median

n	d	γ	α''	α'	n	d	γ	α''	α'
3	1	.750	.250	.125	20	4	.997	.003	.001
4	1	.875	.125	.062		5	.988	.012	.006
5	1	.938	.062	.031		6	.959	.041	.021
6	1	.969	.031	.016		7	.885	.115	.058
	2	.781	.219	.109	21	5	.993	.007	.004
7	1	.984	.016	.008		6	.973	.027	.013
	2	.875	.125	.063		7	.922	.078	.039
8	1	.992	.008	.004		8	.811	.189	.095
	2	.930	.070	.035	22	5	.996	.004	.002
	3	.711	.289	.145		6	.983	.017	.008
9	1	.996	.004	.002		7	.948	.052	.026
	2	.961	.039	.020		8	.866	.134	.067
	3	.820	.180	.090	23	5	.997	.003	.001
10	1	.998	.002	.001		6	.989	.011	.005
	2	.979	.021	.011		7	.965	.035	.017
	3	.891	.109	.055		8	.907	.093	.047
11	1	.999	.001	.000		9	.790	.210	.105
	2	.998	.012	.006	24	6	.993	.007	.003
	3	.935	.065	.033		7	.977	.023	.011
	4	.773	.227	.113		8	.936	.064	.032
12	2	.994	.006	.003		9	.848	.152	.076
	3	.961	.039	.019	25	6	.996	.004	.002
	4	.854	.146	.073		7	.985	.015	.007
13	2	.997	.003	.002		8	.957	.043	.022
	3	.978	.022	.011		9	.892	.108	.054
	4	.908	.092	.046	26	7	.991	.009	.005
	5	.733	.267	.133		8	.971	.029	.014
14	2	.998	.002	.001		9	.924	.076	.038
	3	.987	.013	.006		10	.831	.169	.084
	4	.943	.057	.029	27	7	.994	.006	.003
	5	.820	.180	.090		8	.981	.019	.010
15	3	.993	.007	.004		9	.948	.052	.026
	4	.965	.035	.018		10	.878	.122	.061
	5	.882	.118	.059	28	7	.996	.004	.002
16	3	.996	.004	.002		8	.987	.013	.006
	4	.979	.021	.011		9	.964	.036	.018
	5	.923	.077	.038		10	.913	.087	.044
	6	.790	.210	.105		11	.815	.185	.092
17	3	.998	.002	.001	29	8	.992	.008	.004
	4	.987	.013	.006		9	.976	.024	.012
	5	.951	.049	.025		10	.939	.061	.031
	6	.857	.143	.072		11	.864	.136	.068
18	4	.992	.008	.004	30	8	.995	.005	.003
	5	.969	.031	.015		9	.984	.016	.008
	6	.904	.096	.048		10	.957	.043	.021
	7	.762	.238	.119		11	.901	.099	.049
19	4	.996	.004	.002		12	.800	.200	.100
	5	.981	.019	.010					
	6	.936	.064	.032					
	7	.833	.167	.084					

Table XI (*Continued*)

n	d	γ	α″	α′	n	d	γ	α″	α′
31	8	.997	.003	.002	41	12	.996	.004	.002
	9	.989	.011	.005		13	.988	.012	.006
	10	.971	.029	.015		14	.972	.028	.014
	11	.929	.071	.035		15	.940	.060	.030
	12	.850	.150	.075		16	.883	.117	.059
32	9	.993	.007	.004	42	13	.992	.008	.004
	10	.980	.020	.010		14	.980	.020	.010
	11	.950	.050	.025		15	.956	.044	.022
	12	.890	.110	.055		16	.912	.088	.044
33	9	.995	.005	.002		17	.836	.164	.082
	10	.986	.014	.007	43	13	.995	.005	.003
	11	.965	.035	.018		14	.986	.014	.007
	12	.920	.080	.040		15	.968	.032	.016
	13	.837	.163	.081		16	.934	.066	.033
34	10	.991	.009	.005		17	.874	.126	.063
	11	.976	.024	.012	44	14	.990	.010	.005
	12	.942	.058	.029		15	.977	.023	.011
	13	.879	.121	.061		16	.951	.049	.024
35	10	.994	.006	.003		17	.904	.096	.048
	11	.983	.017	.008		18	.826	.174	.087
	12	.959	.041	.020	45	14	.993	.007	.003
	13	.910	.090	.045		15	.984	.016	.008
	14	.825	.175	.088		16	.964	.036	.018
36	10	.996	.004	.002		17	.928	.072	.036
	11	.989	.011	.006		18	.865	.135	.068
	12	.971	.029	.014	46	14	.995	.005	.002
	13	.935	.065	.033		15	.989	.011	.006
	14	.868	.132	.066		16	.974	.026	.013
37	11	.992	.008	.004		17	.946	.054	.027
	12	.980	.020	.010		18	.896	.104	.052
	13	.953	.047	.024	47	15	.992	.008	.004
	14	.901	.099	.049		16	.981	.019	.009
	15	.812	.188	.094		17	.960	.040	.020
38	11	.995	.005	.003		18	.921	.079	.039
	12	.986	.014	.007		19	.956	.144	.072
	13	.966	.034	.017	48	15	.994	.006	.003
	14	.927	.073	.036		16	.987	.013	.007
	15	.857	.143	.072		17	.971	.029	.015
39	12	.991	.009	.005		18	.941	.059	.030
	13	.976	.024	.012		19	.889	.111	.056
	14	.947	.053	.027	49	16	.991	.009	.005
	15	.892	.108	.054		17	.979	.021	.011
40	12	.994	.006	.003		18	.956	.044	.022
	13	.983	.017	.008		19	.915	.085	.043
	14	.962	.038	.019		20	.848	.152	.076
	15	.919	.081	.040	50	16	.993	.007	.003
	16	.846	.154	.077		17	.985	.015	.008
						18	.967	.033	.016
						19	.935	.065	.032
						20	.881	.119	.059

γ = confidence coefficient
$\alpha' = \frac{1}{2}(1 - \gamma)$ = one-sided significance level
$\alpha'' = 2\alpha' = 1 - \gamma$ = two-sided significance level

Table XII   *d*-Factors for Wilcoxon Signed Rank Test and Confidence Intervals for the Median

n	d	γ	α''	α'	n	d	γ	α''	α'
3	1	.750	.250	.125	15	16	.992	.008	.004
4	1	.875	.125	.062		17	.990	.010	.005
5	1	.938	.062	.031		26	.952	.048	.024
	2	.875	.125	.063		27	.945	.055	.028
6	1	.969	.031	.016		31	.905	.095	.047
	2	.937	.063	.031		32	.893	.107	.054
	3	.906	.094	.047	16	20	.991	.009	.005
	4	.844	.156	.078		21	.989	.011	.006
7	1	.984	.016	.008		30	.956	.044	.022
	3	.953	.047	.016		31	.949	.051	.025
	4	.922	.078	.039		36	.907	.093	.047
	5	.891	.109	.055		37	.895	.105	.052
8	1	.992	.008	.004	17	24	.991	.009	.005
	2	.984	.016	.008		25	.989	.011	.006
	4	.961	.039	.020		35	.955	.045	.022
	5	.945	.055	.027		36	.949	.051	.025
	6	.922	.078	.039		42	.902	.098	.049
	7	.891	.109	.055		43	.891	.109	.054
9	2	.992	.008	.004	18	28	.991	.009	.005
	3	.988	.012	.006		29	.990	.010	.005
	6	.961	.039	.020		41	.952	.048	.024
	7	.945	.055	.027		42	.946	.054	.027
	9	.902	.098	.049		48	.901	.099	.049
	10	.871	.129	.065		49	.892	.108	.054
10	4	.990	.010	.005	19	33	.991	.009	.005
	5	.986	.014	.007		34	.989	.011	.005
	9	.951	.049	.024		47	.951	.049	.025
	10	.936	.064	.032		48	.945	.055	.027
	11	.916	.084	.042		54	.904	.096	.048
	12	.895	.105	.053		55	.896	.104	.052
11	6	.990	.010	.005	20	38	.991	.009	.005
	7	.986	.014	.007		39	.989	.011	.005
	11	.958	.042	.021		53	.952	.048	.024
	12	.946	.054	.027		54	.947	.053	.027
	14	.917	.083	.042		61	.903	.097	.049
	15	.898	.102	.051		62	.895	.105	.053
12	8	.991	.009	.005	·21	43	.991	.009	.005
	9	.988	.012	.006		44	.990	.010	.005
	14	.958	.042	.021		59	.954	.046	.023
	15	.948	.052	.026		60	.950	.050	.025
	18	.908	.092	.046		68	.904	.096	.048
	19	.890	.110	.055		69	.897	.103	.052
13	10	.992	.008	.004	22	49	.991	.009	.005
	11	.990	.010	.005		50	.990	.010	.005
	18	.952	.048	.024		66	.954	.046	.023
	19	.953	.057	.029		67	.950	.050	.025
	22	.906	.094	.047		76	.902	.098	.049
	23	.890	.110	.055		77	.895	.105	.053
14	13	.991	.009	.004	23	55	.991	.009	.005
	14	.989	.011	.005		56	.990	.010	.005
	22	.951	.049	.025		74	.952	.048	.024
	23	.942	.058	.029		75	.948	.052	.026
	26	.909	.091	.045		84	.902	.098	.049
	27	.896	.104	.052		85	.895	.105	.052

Table XII (*Continued*)

n	d	γ	α″	α′	n	d	γ	α″	α′
24	62	.990	.010	.005	25	69	.990	.010	.005
	63	.989	.011	.005		70	.989	.011	.005
	82	.951	.049	.025		90	.952	.048	.024
	83	.947	.053	.026		91	.948	.052	.026
	92	.905	.095	.048		101	.904	.096	.048
	93	.899	.101	.051		102	.899	.101	.051

For $n > 25$ use $d \doteq \frac{1}{2}[\frac{1}{2}n(n+1) + 1 - z\sqrt{n(n+1)(2n+1)/6}]$, where z is read from Table III.
γ = confidence coefficient
$\alpha' = \frac{1}{2}(1 - \gamma)$ = one-sided significance level
$\alpha'' = 2\alpha' = 1 - \gamma$ = two-sided significance level

Table XIII Critical Values for the Kolmogorov-Smirnov Test of Goodness of fit

Sample Size (n)	Significance Level				
	.20	.15	.10	.05	.01
1	.900	.925	.950	.975	.995
2	.684	.726	.776	.842	.929
3	.565	.597	.642	.708	.829
4	.494	.525	.564	.624	.734
5	.446	.474	.510	.563	.669
6	.410	.436	.470	.521	.618
7	.381	.405	.438	.486	.577
8	.358	.381	.411	.457	.543
9	.339	.360	.388	.432	.514
10	.322	.342	.368	.409	.486
11	.307	.326	.352	.391	.468
12	.295	.313	.338	.375	.450
13	.284	.302	.325	.361	.433
14	.274	.292	.314	.349	.418
15	.266	.283	.304	.338	.404
16	.258	.274	.295	.328	.391
17	.250	.266	.286	.318	.380
18	.244	.259	.278	.309	.370
19	.237	.252	.272	.301	.361
20	.231	.246	.264	.294	.352
25	.21	.22	.24	.264	.32
30	.19	.20	.22	.242	.29
35	.18	.19	.21	.23	.27
Asymptotic Formula:	$\dfrac{1.07}{\sqrt{n}}$	$\dfrac{1.14}{\sqrt{n}}$	$\dfrac{1.22}{\sqrt{n}}$	$\dfrac{1.36}{\sqrt{n}}$	$\dfrac{1.63}{\sqrt{n}}$

Reject the hypothetical distribution $F(x)$ if $D_n = \max|F_n(x) - F(x)|$ exceeds the tabulated value.

Table XIV   Critical values for the Kolmogorov-Smirnov test of $H:F_1(x) = F_2(x)$

Sample size n_1

Sample size n_2	1	2	3	4	5	6	7	8	9	10	12	15
1	*	*	*	*	*	*	*	*	*	*		
	*	*	*	*	*	*	*	*	*	*		
2		*	*	*	*		*	7/8	16/18	9/10		
		*	*	*	*	*	*	*	*	*		
3			*	*	12/15	5/6	18/21	18/24	7/9		9/12	
			*	*				*	8/9		11/12	
4				3/4	16/20	9/12	21/28	6/8	27/36	14/20	8/12	
				*	*	10/12	24/28	7/8	32/36	16/20	10/12	
5					4/5	20/30	25/35	27/40	31/45	7/10		10/15
					4/5	25/30	30/35	32/40	36/45	8/10		11/15
6						4/6	29/42	16/24	12/18	19/30	7/12	
						5/6	35/42	18/24	14/18	22/30	9/12	
7							5/7	35/56	40/63	43/70		
							5/7	42/56	4⁷/63	53/70		
8								5/8	45/72	23/40	14/24	
								6/8	54/72	28/40	16/24	
9									5/9	52/90	20/36	
									6/9	62/90	24/36	
10										6/10		15/30
										7/10		19/30
12											6/12	30/60
											7/12	35/60
15												7/15
												8/15

Reject H_0 if

$$D = \max|F_{n_1}(x) - F_{n_2}(x)|$$

exceeds the tabulated value. The upper value gives a level at most .05 and the lower value gives a level at most .01.

Note: Where * appears, do not reject H at the given level.

Table XV *d*-Factors for Wilcoxon-Mann-Whitney Test and Confidence Intervals for the Shift Parameter Δ

		m = 3				*m* = 4		
	d	γ	α″	α′	*d*	γ	α″	α′
γ = confidence coefficient								
$n = 3$	1	.900	.100	.050				
$\alpha' = \frac{1}{2}(1 - \gamma) =$ one-sided significance level								
$n = 4$	1	.943	.057	.029	1	.971	.029	.014
$\alpha'' = 2\alpha' = 1 - \gamma =$ two-sided significance level	2	.886	.114	.057	2	.943	.057	.029
					3	.886	.114	.057
$n = 5$	1	.964	.036	.018	1	.984	.016	.008
	2	.929	.071	.036	2	.968	.032	.016
	3	.857	.143	.071	3	.937	.063	.032
					4	.889	.111	.056
$n = 6$	1	.976	.024	.012	1	.990	.010	.005
	2	.952	.048	.024	2	.981	.019	.010
	3	.905	.095	.048	3	.962	.038	.019
	4	.833	.167	.083	4	.933	.067	.033
					5	.886	.114	.057
$n = 7$	1	.983	.017	.008	1	.994	.006	.003
	2	.967	.033	.017	2	.988	.012	.006
	3	.933	.067	.033	4	.958	.042	.021
	4	.883	.117	.058	5	.927	.073	.036
					6	.891	.109	.055
$n = 8$	1	.988	.012	.006	2	.992	.008	.004
	3	.952	.048	.024	3	.984	.016	.008
	4	.915	.085	.042	5	.952	.048	.024
	5	.867	.133	.067	6	.927	.073	.036
					7	.891	.109	.055
$n = 9$	1	.991	.009	.005	2	.994	.006	.003
	2	.982	.018	.009	3	.989	.011	.006
	3	.964	.036	.018	5	.966	.034	.017
	4	.936	.064	.032	6	.950	.050	.025
	5	.900	.100	.050	7	.924	.076	.038
					8	.894	.106	.053
$n = 10$	1	.993	.007	.004	3	.992	.008	.004
	2	.986	.014	.007	4	.986	.014	.007
	4	.951	.049	.025	6	.964	.036	.018
	5	.923	.077	.039	7	.946	.054	.027
	6	.888	.112	.056	8	.924	.076	.038
					9	.894	.106	.053
$n = 11$	1	.995	.005	.003	3	.994	.006	.003
	2	.989	.011	.006	4	.990	.010	.005
	4	.962	.038	.019	7	.960	.040	.020
	5	.940	.060	.030	8	.944	.056	.028
	6	.912	.088	.044	9	.922	.078	.039
	7	.874	.126	.063	10	.896	.104	.052
$n = 12$	2	.991	.009	.004	4	.992	.008	.004
	3	.982	.018	.009	5	.987	.013	.007
	5	.952	.048	.024	8	.958	.042	.021
	6	.930	.070	.035	9	.942	.058	.029
	7	.899	.101	.051	10	.922	.078	.039
					11	.897	.103	.052

For sample sizes *m* and *n* beyond the range of this table use $d \doteq \frac{1}{2}[mn + 1 - z\sqrt{mn(m + n + 1)/3}]$, where *z* is read from Table III.

Table XV   (*Continued*)

		$m = 5$				$m = 6$				$m = 7$				$m = 8$		
	d	γ	α''	α'	*d*	γ	α''	α'	*d*	γ	α''	α'	*d*	γ	α''	α'
$n = 5$	1	.992	.008	.004												
	2	.984	.016	.008												
	3	.968	.032	.016												
	4	.944	.056	.028												
	5	.905	.095	.048												
	6	.849	.151	.075												
$n = 6$	2	.991	.009	.004	3	.991	.009	.004								
	3	.983	.017	.009	4	.985	.015	.008								
	4	.970	.030	.015	6	.959	.041	.021								
	5	.948	.052	.026	7	.935	.065	.033								
	6	.918	.082	.041	8	.907	.093	.047								
	7	.874	.126	.063	9	.868	.132	.066								
$n = 7$	2	.995	.005	.003	4	.992	.008	.004	5	.993	.007	.004				
	3	.990	.010	.005	5	.986	.014	.007	6	.989	.011	.006				
	6	.952	.048	.024	7	.965	.035	.018	9	.962	.038	.019				
	7	.927	.073	.037	8	.949	.051	.026	10	.947	.053	.027				
	8	.984	.106	.053	9	.927	.073	.037	12	.903	.097	.049				
					10	.899	.101	.051	13	.872	.128	.064				
$n = 8$	3	.994	.006	.003	5	.992	.008	.004	7	.991	.009	.005	8	.993	.007	.004
	4	.989	.011	.005	6	.987	.013	.006	8	.986	.014	.007	9	.990	.010	.005
	7	.955	.045	.023	9	.957	.043	.021	11	.960	.040	.020	14	.950	.050	.025
	8	.935	.065	.033	10	.941	.059	.030	12	.946	.054	.027	15	.935	.065	.033
	9	.907	.093	.047	11	.919	.081	.041	14	.906	.094	.047	16	.917	.083	.042
	10	.873	.127	.064	12	.892	.108	.054	15	.879	.121	.060	17	.895	.105	.052
$n = 9$	4	.993	.007	.004	6	.992	.008	.004	8	.992	.008	.004	10	.992	.008	.004
	5	.988	.012	.006	7	.988	.012	.006	9	.988	.012	.006	11	.989	.011	.006
	8	.958	.042	.021	11	.950	.050	.025	13	.958	.042	.021	16	.954	.046	.023
	9	.940	.060	.030	12	.934	.066	.033	14	.945	.055	.027	17	.941	.059	.030
	10	.917	.083	.042	13	.912	.088	.044	16	.909	.091	.045	19	.907	.093	.046
	11	.888	.112	.056	14	.887	.113	.057	17	.886	.114	.057	20	.886	.114	.057
$n = 10$	5	.992	.008	.004	7	.993	.007	.004	10	.990	.010	.005	12	.991	.009	.004
	6	.987	.013	.006	8	.989	.011	.006	11	.986	.014	.007	13	.988	.012	.006
	9	.960	.040	.020	12	.958	.042	.021	15	.957	.043	.022	18	.957	.043	.022
	10	.945	.055	.028	13	.944	.056	.028	16	.945	.055	.028	19	.945	.055	.027
	12	.901	.099	.050	15	.907	.093	.047	18	.912	.088	.044	21	.917	.083	.042
	13	.871	.129	.065	16	.882	.118	.059	19	.891	.109	.054	22	.899	.101	.051
$n = 11$	6	.991	.009	.004	8	.993	.007	.004	11	.992	.008	.004	14	.991	.009	.005
	7	.987	.013	.007	9	.990	.010	.005	12	.989	.011	.006	15	.988	.012	.006
	10	.962	.038	.019	14	.952	.048	.024	17	.956	.044	.022	20	.959	.041	.020
	11	.948	.052	.026	15	.938	.062	.031	18	.944	.056	.028	21	.949	.051	.025
	13	.910	.090	.045	17	.902	.098	.049	20	.915	.085	.043	24	.909	.091	.045
	14	.885	.115	.058	18	.878	.122	.061	21	.896	.104	.052	25	.891	.109	.054
$n = 12$	7	.991	.009	.005	10	.990	.010	.005	13	.990	.010	.005	16	.990	.010	.005
	8	.986	.014	.007	11	.987	.013	.007	14	.987	.013	.007	17	.988	.012	.006
	12	.952	.048	.024	15	.959	.041	.021	19	.955	.045	.023	23	.953	.047	.024
	13	.936	.064	.032	16	.947	.053	.026	20	.944	.056	.028	24	.943	.057	.029
	14	.918	.082	.041	18	.917	.083	.042	22	.917	.083	.042	27	.902	.098	.049
	15	.896	.104	.052	19	.898	.102	.051	23	.900	.100	.050	28	.885	.115	.058

Table XV (*Continued*)

		$m = 9$				$m = 10$				$m = 11$				$m = 12$		
	d	γ	α''	α'	**d**	γ	α''	α'	**d**	γ	α''	α'	**d**	γ	α''	α'
$n = 9$	12	.992	.008	.004												
	13	.989	.011	.005												
	18	.960	.040	.020												
	19	.950	.050	.025												
	22	.906	.094	.047												
	23	.887	.113	.057												
$n = 10$	14	.992	.008	.004	17	.991	.009	.005								
	15	.990	.010	.005	18	.989	.011	.006								
	21	.957	.043	.022	24	.957	.043	.022								
	22	.947	.053	.027	25	.948	.052	.026								
	25	.905	.095	.047	28	.911	.089	.045								
	26	.887	.113	.056	29	.895	.105	.053								
$n = 11$	17	.990	.010	.005	19	.992	.008	.004	22	.992	.008	.004				
	18	.988	.012	.006	20	.990	.010	.005	23	.989	.011	.005				
	24	.954	.046	.023	27	.957	.043	.022	31	.953	.047	.024				
	25	.944	.056	.028	28	.949	.051	.026	32	.944	.056	.028				
	28	.905	.095	.048	32	.901	.099	.049	35	.912	.088	.044				
	29	.888	.112	.056	33	.886	.114	.057	36	.899	.101	.051				
$n = 12$	19	.991	.009	.005	22	.991	.009	.005	25	.991	.009	.004	28	.992	.008	.004
	20	.988	.012	.006	23	.989	.011	.006	26	.989	.011	.005	29	.990	.010	.005
	27	.951	.049	.025	30	.957	.043	.021	34	.956	.044	.022	34	.955	.045	.023
	28	.942	.058	.029	31	.950	.050	.025	35	.949	.051	.026	39	.948	.052	.026
	31	.905	.095	.048	35	.907	.093	.047	39	.909	.091	.045	43	.911	.089	.044
	32	.889	.111	.056	36	.893	.107	.054	40	.896	.104	.052	44	.899	.101	.050

Table XVI Probabilities Associated with Values as Large as Observed Values of K in the Kruskal-Wallis One-Way Analysis of Variance by Ranks

\ Samples sizes					Samples sizes				
n_1	n_2	n_3	K	p	n_1	n_2	n_3	K	p
2	1	1	2.7000	.500	4	3	2	6.4444	.008
								6.3000	.011
2	2	1	3.6000	.200				5.4444	.046
								5.4000	.051
2	2	2	4.5714	.067				4.5111	.098
			3.7143	.200				4.4444	.102
3	1	1	3.2000	.300	4	3	3	6.7455	.010
								6.7091	.013
3	2	1	4.2857	.100				5.7909	.046
			3.8571	.133				5.7273	.050
								4.7091	.092
3	2	2	5.3572	.029				4.7000	.101
			4.7143	.048					
			4.5000	.067	4	4	1	6.6667	.010
			4.4643	.105				6.1667	.022
								4.9667	.048
3	3	1	5.1429	.043				4.8667	.054
			4.5714	.100				4.1667	.082
			4.0000	.129				4.0667	.102
3	3	2	6.2500	.011	4	4	2	7.0364	.006
			5.3611	.032				6.8727	.011
			5.1389	.061				5.4545	.046
			4.5556	.100				5.2364	.052
			4.2500	.121				4.5545	.098
								4.4455	.103
3	3	3	7.2000	.004					
			6.4889	.011	4	4	3	7.1439	.010
			5.6889	.029				7.1364	.011
			5.6000	.050				5.5985	.049
			5.0667	.086				5.5758	.051
			4.6222	.100				4.5455	.099
4	1	1	3.5714	.200				4.773	.102
4	2	1	4.8214	.057	4	4	4	7.6538	.008
			4.5000	.076				7.5385	.011
			4.0179	.114				5.6923	.049
4	2	2	6.0000	.014				5.6538	.054
			5.3333	.033				4.6539	.097
			5.1250	.052				4.5001	.104
			4.4583	.100					
			4.1667	.105	5	1	1	3.8571	.143
4	3	1	5.8333	.021	5	2	1	5.2500	.036
			5.2083	.050				5.0000	.048
			5.0000	.057				4.4500	.071
			4.0556	.093				4.2000	.095
			3.8889	.129				4.0500 ·	.119

Table XVI (*Continued*)

n_1	n_2	n_3	K	p	n_1	n_2	n_3	K	p
5	2	2	6.5333	.008				5.6308	.050
			6.1333	.013				4.5487	.099
			5.1600	.034				4.5231	.103
			5.0400	.056					
			4.3733	.090	5	4	4	7.7604	.009
			4.2933	.122				7.7440	.011
								5.6571	.049
5	3	1	6.4000	.012				5.6176	.050
			4.9600	.048				4.6187	.100
			4.8711	.052				4.5527	.102
			4.0178	.095					
			3.8400	.123	5	5	1	7.3091	.009
								6.8364	.011
5	3	2	6.9091	.009				5.1273	.046
			6.8218	.010				4.9091	.053
			5.2509	.049				4.1091	.086
			5.1055	.052				4.0364	.105
			4.6509	.091					
			4.4945	.101	5	5	2	7.3385	.010
								7.2692	.010
5	3	3	7.0788	.009				5.3385	.047
			6.9818	.011				5.2462	.051
			5.6485	.049				4.6231	.097
			5.5152	.051				4.5077	.100
			4.5333	.097					
			4.4121	.109	5	5	3	7.5780	.010
								7.5429	.010
5	4	1	6.9545	.008				5.7055	.046
			6.8400	.011				5.6264	.051
			4.9855	.044				4.5451	.100
			4.8600	.056				4.5363	.102
			3.9873	.098					
			3.9600	.102	5	5	4	7.8229	.010
								7.7914	.010
5	4	2	7.2045	.009				5.6657	.049
			7.1182	.010				5.6429	.050
			5.2727	.049				4.5229	.099
			5.2682	.050				4.5200	.101
			4.5409	.098					
			4.5182	.101	5	5	5	8.0000	.009
								7.9800	.010
5	4	3	7.4449	.010				5.7800	.049
			7.3949	.011				5.6600	.051
			5.6564	.049				4.5600	.100
								4.5000	.102

Table XVII   Critical Values of Friedman Statistic

t	b	.90	.95	.975	.99	.995
3	3	6.000	6.000	—	—	—
3	4	6.000	6.500	8.000	8.000	8.000
3	5	5.200	6.400	7.600	8.400	10.000
3	6	5.333	7.000	8.333	9.000	10.333
3	7	5.429	7.143	7.714	8.857	10.286
3	8	5.250	6.250	7.750	9.000	9.750
3	9	5.556	6.222	8.000	8.667	10.667
4	2	6.000	6.000	—	—	—
4	3	6.600	7.400	8.200	9.000	9.000
4	4	6.300	7.800	8.400	9.600	10.200

Acknowledgements for Table

Table IV. Reprinted from *Biometrika Tables for Statisticians*, Vol. I, by E. S. Pearson and H. O. Hartley, 1970. By permission of *Biometrika* Trustees.

Table V. Reprinted from *Biometrika Tables for Statisticians*, Vol. I, by E. S. Pearson and H. O. Hartley, 1970. By permission of *Biometrika* Trustees.

Table VI. Reprinted from *Biometrika Tables for Statisticians*, Vol. I, by E. S. Pearson and H. O. Hartley, 1970. By permission of *Biometrika* Trustees.

Table VII. *A Million Random Digits with 100,000 Normal Deviates*, by The Rand Corporation. New York: The Free Press, 1955. Copyright 1955 and 1983 by the Rand Corporation. Reprinted from pp. 115, 258, 262, and 360. Used by permission.

Table VIII. Reprinted from *Biometrika Tables for Statisticians*, Vol. I, by E. S. Pearson and H. O. Hartley, 1970. By permission of *Biometrika* Trustees.

Table IX. Adapted from H. Leon Harter, Tables of range and studentized range, *Annals of Mathematical Statistics*, 1960, 31, 1122–47. By permission of author and publisher.

Table X. Adapted from E. G. Olds, Distribution of sums of sq of rank differences for small number of individuals, *Annals of Mathematical Statistics*, 1938, 9, 133–148. By permission of author and publisher.

Table XI. Reprinted from Gottfried Noether, *Introduction to Statistics: A Fresh Approach*. Boston: Houghton Mifflin Company, 1971. Used by permission.

Table XII. Reprinted from Gottfried Noether, *Introduction to Statistics: A Fresh Approach*, Boston: Houghton Mifflin Company, 1971. Used by permission.

Table XIII. Adapted from Frank Massey, Jr., The Kolmogorov-Smirnov test for goodness of fit, *Journal of American Statistical Association*, 1951, 46, 68–78. Used by permission.

Table XIV. Adapted from F. J. Massey, Jr., Distribution table for the deviation between two sample cumulatives, *Annals of Mathematical Statistics*, 1952, 23, 435–441. By permission of author and publisher.

Table XV. Reprinted from Gottfried Noether, *Introduction to Statistics: A Fresh Approach*. Boston: Houghton Mifflin Company, 1971. Used by permission.

Table XVI. Adapted from W. H. Kruskal and W. A. Wallis, Use of ranks in one-criterion variance analysis, *Journal of American Statistical Association*, 1952, 47, 581–621. Used by permission.

Table XVII. Adapted from M. Friedman, The use of ranks to avoid the assumption of normality implicit in the analysis of variance, *Journal of American Statistical Association*, 1937, 52, 675–701. Used by permission.

Index